Counting Money Correctly

by

Lonnie Joe Noyes

First published by AuthorHouse 04/05/04

ISBN: 1-4140-2023-6 (e-book)
ISBN: 1-4184-0559-0 (Paperback)

Library of Congress Control Number: 2003099925

This book is printed on acid free paper.

Printed in the United States of America
Bloomington, IN

TABLE OF CONTENTS

Mathematical Solutions Without Answers .... 10 - 37

Mathematical Solutions With Answers .... 38 - 65

PENNIES-------THE ONE CENT COIN

1. One (1), Penny is One - Cent.
2. There are Five (5), Pennies in One Nickel.
3. There are Ten (10), Pennies in One Dime.
4. There are Twenty-Five (25), Pennies in One Quarter.
5. There are Fifty (50), Pennies in a Half-Dollar.
6. There are One-Hundred (100), Pennies in One-Dollar.

NICKELS-------THE FIVE CENTS COIN

1. One (1), Nickel is Five - Cents.
2. .There are Five (5), Pennies in One Nickels.
3. There are Two (2), Nickels in One Dime.
4. There are Five (5), Nickels in One Quarter.
5. There are Ten (10), Nickels in a Half-Dollar.
6. There are Twenty (20), Nickels in One Dollar.

DIMES-------THE TEN CENTS COIN

1. One (1), Dime is Ten - Cents.
2. There are Ten (10), Pennies in One Dime.
3. There are Two (2), Nickels in One Dime.
4. There are Five (5), Dimes in a Half-Dollar.
5. There are Ten (10), Dimes in a One Dollar Bill.

QUARTERS-------THE TWENTY-FIVE CENTS COIN

1. One (1), Quarter is Twenty-Five Cents.
2. There are Twenty-Five (25), Pennies in One Quarter.
3. There are Five (5), Nickels in One Quarter.
4. There are Three (3), Nickels and One (1), Dime in One Quarter.
5. There are Two (2), Dimes and One (1), Nickel in One Quarter.
6. There are Four (4), Quarters in One Dollar.

COUNTING MONEY CORRECTLY REVIEW

1.A Penney is a One (1) – Cent Coin.
2. A Nickel is a Five (5) - Cents Coin.
3. A Dime is a Ten (10) - Cents Coin.
4. A Quarter is a Twenty-Five (25) - Cents Coin.
5. A Half-Dollar is a Fifty (50) - Cents Coin.
6. A Silver-Dollar is a One-Hundred (100) - Cents Coin. Also a Gold Dollar and a Paper One-Dollar Bill has One-Hundred (100) - Cents.
7. There are Two Nickels in One Dime.
8. One Dime is Ten (10) - Cents.
9. There are One Nickel and One Dime in Fifteen - (15) - Cents.
10. There are Two Dimes in Twenty (20) - Cents.
11. A Quarter is a Twenty-five - (25) - Cents Coin.
12. Two Dimes and One Nickel is Twenty-Five (25) - Cents.
13 One Dime and Three Nickels is Twenty-Five (25) - Cents.
14. Also, Five Nickels is Twenty-Five (25) - Cents.
15. A Quarter and a Nickel is Thirty (30) - Cents.
16. A Quarter and a Dime is Thirty-Five (35) - Cents.
17. Also, a Quarter and Two Nickels is Thirty-Five (35) - Cents.
18. A Quarter, a Nickel and a Dime is Forty (40) - Cents.

19. A Quarter, Two Nickels and a Dime is Forty-Five (45) - Cents.
20. A Quarter and Two Dimes is Forty-Five (45) - Cents.
21. Also, a Quarter and Four Nickels is Forty-Five (45) - Cents.
22. A Half-Dollar is a Fifty (50) - Cents Coins.
23. Two Quarters is Fifty (50) - Cents.
24. One Quarter, One Nickel and Two Dimes is Fifty (50) - Cents.
25. One Quarter, Three Nickels, and One Dime is Fifty (50) - Cents.
26. Also, One Quarter and Five Nickels is Fifty (50) - Cents.
27. Two Quarters and One Nickel is Fifty-Five (55) - Cents.
28. Two Quarters and One Dime is Sixty (60) - Cents.
29. Also, Two Quarters and Two Nickels is Sixty (60)- -Cents.
30. Two Quarters, One Nickel, and One Dime is Sixty-Five (65) - Cents.
31. Two Quarters, and Two Dimes is Seventy (70) - Cents.
32. Two Quarters, One Dime, and Two Nickels is Seventy (70) - Cents.
33. Also, Two Quarters and Four Nickels is Seventy (70) - Cents.
34. Three Quarters is Seventy-Five (75) - Cents.
35. One Quarter and One Half-Dollar is Seventy-Five (75) - Cents.
36. One Quarter and Five Dimes is Seventy-Five (75) - Cents.
37. One Quarter and Ten Nickels is Seventy-Five (75) - Cents.
38. Two Quarters, Two Dimes, and One Nickel is Seventy-Five (75) - Cents.
39. Two Quarters, One Dime, and Three Nickels is Seventy-Five (75) - Cents.
40. Also Two Quarters and Five Nickels is Seventy-Five (75) - Cents.
41. Three Quarters and One Nickel is Eighty (80) - Cents.
42. Three Quarters and One Dime is Eighty-Five (85) - Cents.
43. Also Three Quarters and Two Nickels is Eighty-Five (85) - Cents.
44. Three Quarters, One Nickel, and One Dime is Ninety (90) - Cents.
45. Three Quarters and Two Dimes is Ninety-Five (95) - Cents.
46. Three Quarters, One Dime, and Two Nickels is Ninety-Five (95) - Cents.
47. Also Three Quarters and Four Nickels is Ninety-Five (95) - Cents.

THE ONE DOLLAR BILL

1. There are Two (2), Half-Dollars in One Dollar (100) - Cents.
2. There are One (1), Half-Dollar and Two (2), Quarters in One Dollar (100) - Cents.
3. There are One Half-Dollar, One Quarter, One Dime, and Three Nickels in One Dollar (100) - Cents.
4. There are One (1), Half-Dollar, One (1), Quarter, Two (2), Dimes, and One (1), Nickel in One Dollar (100) – Cents.
5. There are Four (4), Quarters in One Dollar.
6. There are Three (3), Quarters, Two (2), Dime, and One (1), Nickel in One Dollar (100) - Cents.
7. There are Three (3), Quarters, One (1), Dimes, and Three (3), Nickels in One Dollar (100) - Cents.
8. There areTwo (2), Quarters and Fives (5), Dimes in One Dollar (100) - Cents.
9. There are Two (2), Quarters and Ten- (10), Nickels in One Dollar (100) - Cents.
10. There are One (1), Quarter, Seven (7), Dimes and One (1), Nickel in One Dollar (100) - Cents.
11. There are One (1), Quarter and Fifteen (15), Nickels in One Dollar (100) - Cents.
12. There are Ten (10), Dimes in One Dollar (100) - Cents.
13. Also Five (5), Dimes and Ten (10), Nickels in One Dollar Bill (100) - Cents.
14. There are Twenty (20), Nickels in One Dollar.
15. There are One-Hundred (100), Pennies in One Dollar.

THE FIVE DOLLAR BILL

1. There are Five – (5), One-Dollar Bills in Five-Dollars.
2. There are Ten (10), Half-Dollars in Five-Dollars.
3. There are Twenty – (20), Quarters Five-Dollars.
4. There are Fifty – (50), Dimes in Five-Dollars.
5. There are One-Hundred - (100), Nickels in Five-Dollars.
6. There are Five-Hundred- (500), Pennies in Five-Dollars.

THE TEN DOLLAR BILL

1. Two – (2) Five-Dollar Bills in Ten-Dollars.
2. There are Ten - (10), One-Dollar Bills in Ten Dollars.
3. There are Forty – (40), Quarters in Ten-Dollars.
4. There are One-Hundred Dimes in Ten-Dollars.
5. There are Two-Hundred Nickels in Ten-Dollars.
6. There are One-Thousand Pennies in Ten-Dollars.

THE TWENTY DOLLAR BILL

1. There are Two - (2), Ten-Dollar Bills in Twenty Dollars.
2. There are Four- (4), Five-Dollar Bills in Twenty Dollars.
3. There are One (1), Ten-Dollar Bill and Two - (2), Five-Dollar Bills in Twenty-Dollars.
4. There are One - (1), Ten-Dollar Bill, One - (1), Five-Dollar Bill and Five –(5), One-(1),Dollar Bills in Twenty-Dollars.
5. There are Three - (3), Five-Dollar Bills and Five – (5), One-Dollar Bills in Twenty-Dollars.
6. There are One - (1), Ten-Dollar Bill and Ten - (10), One-Dollar Bills in Twenty-Dollars.

7. There are One - (1), Five-Dollar Bill and Fifteen - (15), One-Dollar Bills in Twenty-Dollars.

8. There are Twenty - (20), One-Dollar Bills in Twenty-Dollars.

9. There are Eighty - (80), Quarters in Twenty-Dollars.

10. There are Two-Hundred - (200), Dimes in Twenty-Dollars.

11. There are Four-Hundred – (400), Nickels in Twenty-Dollars.

12. There are Two-Thousand-(2,000), Pennies in Twenty-Dollars.

THE FIFTY DOLLAR BILL

1. There are Two - (2), Twenty-Dollar Bills and One - (1), Ten-Dollar Bill in Fifty-Dollars.

2. There are One - (1), Twenty-Dollar Bill and Three – (3), Ten-Dollar Bills in Fifty-Dollars.

3. There are Five – (5), Ten-Dollar Bills in Fifty-Dollars.

4. There are Four – (4), Ten-Dollar Bills and Two – (2), Five-Dollar Bills in Fifty-Dollars.

5. There are Three – (3), Ten-Dollar Bills and Four – (4), Five-Dollar Bills in Fifty-Dollars.

6. There are Two – (2), Ten-Dollar Bills and Six – (6), Five-Dollar Bills in Fifty-Dollars.

7. There are One – (1), Ten-Dollar Bill and Eight – (8), Five-Dollar Bills in Fifty-Dollars.

8. There are Ten – (10), Five-Dollar Bills in Fifty-Dollars.

9. There are Fifty – (50), One-Dollar Bills in Fifty-Dollars.

10. There are Two-Hundred (200), Quarters in Fifty-Dollars.

11. There are Five-Hundred (500), Dimes in Fifty-Dollars.

12. There are One-Thousand (1,000), Nickels in Fifty-Dollars.

13. There are Five-Thousand (5,000), Pennies in Fifty-Dollars.

THE HUNDRED DOLLAR BILL

1.There are Two – (2), Fifty-Dollar Bills in One-Hundred Dollars.

2. There are One – (1), Fifty-Dollar Bill, Two – (2), Twenty-Dollar Bill and One – (1), Ten-Dollar Bill in One-Hundred-Dollars.

3.There are One – (1), Fifty-Dollar Bill, One – (1), Twenty-Dollar Bill and Three – (3), Ten-Dollar Bills in One-Hundred Dollars.

4. There are One – (1), Fifty-Dollar Bill and Five – (5), Ten-Dollar Bills in One-Hundred Dollars.

5. There are Five – (5), Twenty-Dollar Bills in One-Hundred Dollars.

6. There are Four – (4), Twenty-Dollar Bills and Two – (2), Ten-Dollar Bills in One-Hundred Dollars.

7. There are Three – (3), Twenty-Dollar Bills and Four – (4), Ten-Dollar Bills in a One-Hundred Dollar Bill.

8. There are Two – (2), Twenty-Dollars Bills and Six – (6), Ten-Dollar Bills in One-Hundred Dollars.

9. There are One – (1), Twenty-Dollar Bill and Eight – (8), Ten-Dollar Bills in One-Hundred Dollars.

10. There are Ten – (10), Ten-Dollar Bills in One-Hundred Dollars.

11. There are Five – (5), Ten-Dollar Bills and Ten – (10), Five-Dollars Bills in One-Hundred Dollars.

12. There are Twenty- (20), Five-Dollar Bills in One-Hundred Dollars.

13. There are Four-Hundred (400), Quarters in One-Hundred Dollars.

14. There are One-Thousand (1000), Dimes in One-Hundred Dollars.

15. There are Two-Thousand (2000), Nickels in One-Hundred Dollars.

16. There are Ten-Thousand (10,000), Pennies in One-Hundred Dollars.

THE FIVE-HUNDRED DOLLAR BILL

1. There are Five (5), One-Hundred Dollar Bills in Five-Hundred Dollars.
2. There are Ten (10), Fifty-Dollar Bills in Five-Hundred Dollars.
3. There are Twenty-Five (25), Twenty-Dollar Bills in Five-Hundred Dollars.
4. There are Fifty (50) Ten-Dollar Bills in Five-Hundred Dollars.
5. There are One-Hundred (100), Five-Dollar Bills in Five-Hundred Dollars.
6. There are Five-Hundred (500), One-Dollar Bills in Five-Hundred Dollars.
7. There are Two-Thousand (2,000), Quarters in Five-Hundred Dollars.
8. There are Five-Thousand (5,000), Dimes in Five-Hundred Dollars.
9. There are Ten-Thousand (10,000), Nickels in Five_Hundred Dollars.
10. There are Fifty-Thousand (50,000), Pennies in Five-Hundred Dollars.

THE ONE-THOUSAND DOLLAR BILL

1. There are Two (2), Five-Hundred Dollar Bills in One Thousand Dollars.
2. There are Ten (10), One-Hundred Dollar Bills in One Thousand Dollars.
3. There are Twenty (20), Fifty-Dollar Bills in Thousand Dollars.
4. There are Fifty (50), Twenty-Dollar Bills in One Thousand Dollars.
5. There are One-Hundred (100), Ten-Dollar Bills in One Thousand Dollars.
6. There are Two-Hundred (200), Five-Dollar Bills in One Thousand Dollars.
7. There are One-Thousand (1,000), One-Dollar Bills in One Thousand Dollars.
8. There are Four-Thousand (4,000), Quarters in One Thousand Dollars.
9. There are Ten-Thousand (10,000), Dimes in One Thousand Dollars.
10. There are Twenty-Thousand (20,000), Nickels in One Thousand Dollars.
11. There are One Hundred-Thousand (100,000), Pennies in One Thousand Dollars.

THE MILLION DOLLAR BILL

1.There are One-Thousand (1,000), Thousand-Dollar Bills in One-Million Dollars.

2. There are Two-Thousand (2,000), Five-Hundred Dollar Bills in One-Million Dollars.

3. There are Ten-Thousand (10,000), One-Hundred Dollar Bills in One-Million Dollars.

4. There are Twenty-Thousand (20,000), Fifty-Dollar Bills in One-Million Dollars.

5. There are Fifty-Thousand (50,000), Twenty-Dollar Bills in One-Million Dollar Bill.

6. There are One-Hundred Thousand (100,000), Ten-Dollar Bills in One-Million Dollar Bill.

7. There are Two-Hundred Thousand (200,000) Five-Dollar Bills in One-Million Dollars.

8. There are One-Million (1,000,000), One-Dollar Bills in One-Million Dollars.

9. There are Four-Million (4,000,000), Quarters in One-Million Dollars.

10. There are Ten-Million (10,000,000), Dimes in One-Million Dollars.

11. There are Twenty-Million (20,000,000), Nickels in One-Million Dollars.

12. There are One-Hundred Million (100,000,000), Pennies in One-Million Dollars.

ADDING NUMBERS

0 + 1 =	0 + 2 =	0 + 3 =	0 + 4 =	0 + 5 =
1 + 1 =	2 + 1 =	3 + 1 =	4 + 1 =	5 + 1 =
1 + 2 =	2 + 2 =	3 + 2 =	4 + 2 =	5 + 2 =
1 + 3 =	2 + 3 =	3 + 3 =	4 + 3 =	5 + 3 =
1 + 4 =	2 + 4 =	3 + 4 =	4 + 4 =	5 + 4 =
1 + 5 =	2 + 5 =	3 + 5 =	4 + 5 =	5 + 5 =
1 + 6 =	2 + 6 =	3 + 6 =	4 + 6 =	5 + 6 =
1 + 7 =	2 + 7 =	3 + 7 =	4 + 7 =	5 + 7 =
1 + 8 =	2 + 8 =	3 + 8 =	4 + 8 =	5 + 8 =
1 + 9 =	2 + 9 =	3 + 9 =	4 + 9 =	5 + 9 =
1 + 10 =	2 + 10 =	3 +10 =	4 + 10 =	5 + 10 =
1 + 11 =	2 + 11 =	3 +11 =	4 + 11 =	5 + 11=
1 + 12 =	2 + 12 =	3 +12 =	4 + 12 =	5 + 12 =
1 + 13 =	2 + 13 =	3 +13 =	4 + 13 =	5 + 13 =
1 + 14 =	2 + 14 =	3 + 14 =	4 + 14 =	5 + 14 =
1 + 15 =	2 + 15 =	3 + 15 =	4 + 15 =	5 + 15 =
1 + 16 =	2 + 16 =	3 + 16 =	4 + 16 =	5 + 16 =
1 + 17 =	2 + 17 =	3 + 17 =	4 + 17 =	5 + 17 =
1 + 18 =	2 + 18 =	3 + 18 =	4 + 18 =	5 + 18 =
1 + 19 =	2 + 19 =	3 + 19 =	4 + 19 =	5 + 19 =

1 + 20 =	2 +20 =	3 + 20 =	4 + 20 =	5 + 20 =
1 + 21=	2+ 21 =	3 + 21 =	4 + 21 =	5 + 21 =
1 + 22 =	2 +22 =	3 + 22 =	4 + 22 =	5 + 22 =
1 + 23 =	2 +23 =	3 + 23 =	4 + 23 =	5 + 23 =
1 + 24 =	2 + 24 =	3 + 24 =	4 + 24 =	5 + 24 =
1 + 25 =	2 + 25 =	3 + 25 =	4 + 25 =	5 + 25 =

6 + 0 =	7 + 0 =	8 + 0 =	9 + 0 =	10 + 0 =
6 + 1 =	7 + 1 =	8 + 1 =	9 +1 =	10 + 1 =
6 + 2 =	7 + 2 =	8 + 2 =	9 + 2 =	10 + 2 =
6 + 3 =	7 + 3 =	8 + 3 =	9 + 3 =	10 + 3 =
6 + 4 =	7 + 4 =	8 + 4 =	9 + 4 =	10 + 4 =
6 + 5 =	7 + 5 =	8 + 5 =	9 + 5 =	10 + 5 =
6 + 6 =	7 + 6 =	8 + 6 =	9 + 6 =	10 + 6 =
6 + 7 =	7 + 7 =	8 + 7 =	9 + 7 =	10 + 7 =
6 + 8 =	7 + 8 =	8 + 8 =	9 + 8 =	10 + 8 =
6 + 9 =	7 + 9 =	8 + 9 =	9 + 9 =	10 + 9 =
6 + 10 =	7 + 10 =	8 + 10 =	9 + 10 =	10 + 10 =
6 + 11 =	7 + 11 =	8 + 11 =	9 + 11 =	10 + 11 =
6 + 12 =	7 + 12 =	8 + 12 =	9 + 12 =	10 + 12 =
6 + 13 =	7 + 13 =	8 + 13 =	9 + 13 =	10 + 13 =
6 + 14 =	7 + 14 =	8 + 14 =	9 + 14 =	10 + 14 =
6 + 15 =	7 + 15 =	8 + 15 =	9 + 15 =	10 + 15 =
6 + 16 =	7 + 16 =	8 + 16 =	9 + 16 =	10 + 16 =
6 + 17 =	7 + 17 =	8 + 17 =	9 + 17 =	10 + 17 =
6 + 18 =	7 + 18 =	8 + 18 =	9 + 18 =	10 + 18 =
6 + 19 =	7 + 19 =	8 + 19 =	9 + 19 =	10 + 19 =
6 + 20 =	7 + 20 =	8 + 20 =	9 + 20 =	10 + 20 =
6 + 21 =	7 + 21 =	8 + 21 =	9 + 21 =	10 + 21 =
6 + 22 =	7 + 22 =	8 + 22 =	9 + 22 =	10 + 22 =

6 + 23 =	7 + 23 =	8 + 23 =	9 + 23 =	10 + 23 =
6 + 24 =	7 + 24 =	8 + 24 =	9 + 24 =	10 + 24 =
6 + 25 =	7 + 25 =	8 + 25 =	9 + 25 =	10 + 25 =

11 + 0 =	12 + 0 =	13 + 0 =	14 + 0 =	15 + 0 =
11 + 1 =	12 + 1 =	13 + 1 =	14 + 1 =	15 + 1 =
11 + 2 =	12 + 2 =	13 + 2 =	14 + 2 =	15 + 2 =
11 + 3 =	12 + 3 =	13 + 3 =	14 + 3 =	15 + 3 =
11 + 4 =	12 + 4 =	13 + 4 =	14 + 4 =	15 + 4 =
11 + 5 =	12 + 5 =	13 + 5 =	14 + 5 =	15 + 5 =
11 + 6 =	12 + 6 =	13 + 6 =	14 + 6 =	15 + 6 =
11 + 7 =	12 + 7 =	13 + 7 =	14 + 7 =	15 + 7 =
11 + 8 =	12 + 8 =	13 + 8 =	14 + 8 =	15 + 8 =
11 + 9 =	12 + 9 =	13 + 9 =	14 + 9 =	15 + 9 =
11 + 10 =	12 + 10 =	13 + 10 =	14 + 10 =	15 + 10 =
11 + 11 =	12 + 11 =	13 + 11 =	14 + 11 =	15 + 11 =
11 + 12 =	12 + 12 =	13 + 12 =	14 + 12 =	15 + 12 =
11+ 13 =	12 + 13 =	13 + 13 =	14 + 13 =	15 + 13 =
11+ 14 =	12 + 14 =	13 + 14 =	14 + 14 =	15 + 14 =
11+ 15 =	12 + 15 =	13 + 15 =	14 + 15 =	15 + 15 =
11 + 16 =	12 + 16 =	13 + 16 =	14 + 16 =	15 + 16 =
11 + 17 =	12 + 17 =	13 + 17 =	14 + 17 =	15 + 17 =
11 + 18 =	12 + 18 =	13 + 18 =	14 + 18 =	15 + 18 =
11 + 19 =	12 + 19 =	13 + 19 =	14 +19 =	15 + 19 =
11 + 20 =	12 + 20 =	13 + 20 =	14 + 20 =	15 + 20 =
11 + 21 =	12 + 21 =	13 +21 =	14 + 21 =	15 + 21 =
11 + 22 =	12 + 22 =	13 + 22 =	14 + 22 =	15 + 22 =
11 + 23 =	12 + 23 =	13 + 23 =	14 + 23 =	15 + 23 =
11 + 24 =	12 + 24 =	13 + 24 =	14 + 24 =	15 + 24 =
11 + 25 =	12 + 25 =	13 + 25 =	14 + 25 =	15 + 25 =

16 + 0 =	17 + 0 =	18 + 0 =	19 + 0 =	20 + 0 =
16 + 1 =	17 + 1 =	18 + 1 =	19 + 1 =	20 + 1 =
16 + 2 =	17 + 2 =	18 + 2 =	19 + 2 =	20 + 2 =
16 + 3 =	17 + 3 =	18 + 3 =	19 + 3 =	20 + 3 =
16 + 4 =	17 + 4 =	18 + 4 =	19 + 4 =	20 + 4 =
16 + 5 =	17 + 5 =	18 + 5 =	19 + 5 =	20 + 5 =
16 + 6 =	17 + 6 =	18 + 6 =	19 +6 =	20 + 6 =
16 + 7 =	17 + 7 =	18 +7 =	19 + 7 =	20 + 7 =
16 + 8 =	17 + 8 =	18 + 8 =	19 + 8 =	20 + 8 =
16 + 9 =	17 + 9 =	18 + 9 =	19 + 9 =	20 + 9 =
16 + 10 =	17 + 10 =	18 + 10 =	19 + 10 =	20 + 10 =
16 + 11 =	17 + 11 =	18 + 11 =	19 + 11 =	20 + 11 =
16 + 12 =	17 + 12 =	18 + 12 =	19 + 12 =	20 + 12 =
16 + 13 =	17 + 13 =	18 +13 =	19 +13 =	20 + 13 =
16 + 14 =	17 + 14 =	18 +14 =	19 + 14 =	20 + 14 =
16 + 15 =	17 + 15 =	18 +15 =	19 + 15 =	20 + 15 =
16 + 16 =	17 + 16 =	18 + 16 =	19 +16 =	20 + 16 =
16 + 17 =	17 + 17 =	18 + 17 =	19 + 17 =	20 + 17 =
16 +18 =	17 + 18 =	18 + 18 =	19 + 18 =	20 + 18 =
16 +19 =	17 + 19 =	18 + 19 =	19 + 19 =	20 + 19 =
16 + 20 =	17 + 20 =	18 + 20 =	19 + 20 =	20 + 20 =
16 + 21 =	17 + 21 =	18 + 21 =	19 + 21 =	20 + 21 =
16 + 22 =	17 + 22 =	18 + 22 =	19 + 22 =	20 + 22 =
16 + 23 =	17 + 23 =	18 + 23 =	19 + 23 =	20 + 23 =
16 + 24 =	17 + 24 =	18 + 24 =	19 + 24 =	20 + 24 =
16 + 25 =	17 + 25 =	18 + 25 =	19 + 25 =	20 + 25 =

21 + 0 =	22 + 0 =	23 + 0 =	24 + 0 =	25 + 0 =
21 + 1 =	22 + 1 =	23 + 1 =	24 + 1 =	25 + 1 =

21 + 2 =	22 + 2 =	23 + 2 =	24 + 2 =	25 + 2 =
21 + 3 =	22 + 3 =	23 + 3 =	24 + 3 =	25 + 3 =
21 + 4 =	22 + 4 =	23 + 4 =	24 + 4 =	25 + 4 =
21 + 5 =	22 + 5 =	23 + 5 =	24 + 5 =	25 + 5 =
21 + 6 =	22 + 6 =	23 + 6 =	24 + 6 =	25 + 6 =
21 + 7 =	22 + 7 =	23 + 7 =	24 + 7 =	25 + 7 =
21 + 8 =	22 + 8 =	23 + 8 =	24 + 8 =	25 + 8 =
21 + 9 =	22 + 9 =	23 + 9 =	24 + 9 =	25 + 9 =
21 + 10 =	22 + 10 =	23 +10 =	24 +10 =	25 + 10 =
21 + 11 =	22 + 11=	23 + 11=	24 + 11=	25 + 11=
21 + 12 =	22 + 12 =	23 +12 =	24 + 12 =	25 + 12 =
21 + 13 =	22 + 13 =	23 + 13 =	24 + 13 =	25 + 13 =
21 + 14 =	22 + 14 =	23 + 14 =	24 + 14 =	25 + 14 =
21 + 15 =	22 + 15 =	23 + 15 =	24 + 15 =	25 + 15 =
21 + 16 =	22 + 16 =	23 + 16 =	24 + 16 =	25 + 16 =
21 + 17 =	22 + 17 =	23 + 17=	24 + 17 =	25 + 17 =
21 + 18 =	22 + 18 =	23 + 18 =	24 + 18 =	25 + 18 =
21 + 19 =	22 + 19 =	23 + 19 =	24 + 19 =	25 + 19 =
21 + 20 =	22 + 20 =	23 + 20 =	24 + 20 =	25 + 20 =
21 + 21 =	22 + 21 =	23 + 21 =	24 + 21 =	25 +21 =
21 + 22 =	22 + 22 =	23 + 22 =	24 + 22 =	25 + 22 =
21 + 23 =	22 + 23 =	23 + 23 =	24 + 23 =	25 + 23 =
21 + 24 =	22 + 24 =	23 + 24 =	24 + 24 =	25 + 24 =
21 + 25 =	22 + 25 =	23 + 25 =	24 + 25 =	25 + 25 =

SUBTRACTING NUMBERS

1- 0 =	2 - 0 =	3 - 0 =	4 - 0 =	5 – 0 =
1 - 1 =	2 - 1 =	3 - 1 =	4 - 3 =	5 - 1 =
2 - 1 =	2 - 2 =	3 – 2 =	4 – 2 =	5 - 2 =
3 - 1 =	3 - 2 =	3 – 3 =	4 – 3 =	5 - 3 =
4 - 1 =	4 - 2 =	4 - 3 =	4 – 4 =	5 – 4 =
5 – 1 =	5 – 2 =	5 – 3 =	5 - 4 =	5 - 5 =
6 – 1 =	6 - 2 =	6 – 3 =	6 – 4 =	6 – 5 =
7 – 1 =	7 – 2 =	7 – 3 =	7 – 4 =	7 – 5 =
8 – 1 =	8 - 2 =	8 - 3 =	8 – 4 =	8 - 5 =
9 - 1 =	9 - 2 =	9 - 3 =	9 - 4 =	9 - 5 =
10 – 1 =	10 - 2 =	10 - 3 =	10 – 4 =	10 - 5 =
10 – 1 =	10 - 2 =	10 - 3 =	10 – 4 =	10 - 5 =
11 – 1 =	11 - 2 =	11 - 3 =	11 - 4 =	11 - 5 =
12 – 1 =	12 - 2 =	12 - 3 =	12 - 4 =	12 - 5 =
13 – 1 =	13 – 2 =	13 - 3 =	13 - 4 =	13 - 5 =
14 – 1 =	14 - 2 =	14 - 3 =	14 - 4 =	14 - 5 =
15 - 1 =	15 - 2 =	15 - 3 =	15 – 4 =	15 - 5 =
16 - 1 =	16 - 2 =	16 - 3 =	16 - 4 =	16 - 5 =
17 – 1 =	17 – 2 =	17 – 3 =	17 – 4 =	17 – 5 =
18 – 1 =	18 – 2 =	18 – 3 =	18 – 4 =	18 – 5 =

19 - 1 =	19 - 2 =	19 - 3 =	19 - 4 =	19 - 5 =
20 - 1 =	20 - 2 =	20 - 3 =	20 - 4 =	20 - 5 =
21 - 1 =	21- 2 =	21 - 3 =	21- 4 =	21- 5 =
22 - 1 =	22 -2 =	22 - 3 =	22 - 4 =	22 - 5 =
23 – 1 =	23 - 2 =	23 - 3 =	23 - 4 =	23 - 5 =
24 - 1 =	24 - 2 =	24 - 3 =	24 - 4 =	24 - 5 =
25 - 1 =	25 - 2 =	25 - 3 =	25 - 4 =	25 - 5 =

6 - 0 =	7 - 0 =	8 - 0 =	9 - 0 =	10 - 0 =
6 - 1 =	7 - 1 =	8 - 1 =	9 - 1 =	10 - 1 =
6 - 2 =	7 - 2 =	8 - 2 =	9 - 2 =	10 - 2 =
6 - 3 =	7 - 3 =	8 - 3 =	9 - 3 =	10 - 3 =
6- 4 =	7 - 4 =	8 - 4 =	9 - 4 =	10 - 4 =
6 - 5 =	7- 5 =	8 - 5 =	9 - 5 =	10 - 5 =
6 - 6 =	7 - 6 =	8 - 6 =	9 - 6 =	10 - 6 =
7 – 6 =	7 – 7 =	8 – 7 =	9 – 7 =	10 – 7 =
8 – 6 =	8 - 7 =	8 – 8 =	9 - 8 =	10 – 8 =
9 - 6 =	9 – 7 =	9 - 8 =	9 - 9 =	10 – 9 =
10 - 6 =	10 - 7 =	10 - 8 =	10 - 9 =	10 - 10 =
11- 6 =	11- 7 =	11 - 8 =	11- 9 =	11 - 10 =
12 - 6 =	12 - 7 =	12 - 8 =	12 - 9 =	12 - 10 =
13 - 6 =	13 - 7 =	13 - 8 =	13 - 9 =	13 -10 =
14 - 6 =	14 - 7 =	14 - 8 =	14 - 9 =	14 - 10 =
15 - 6 =	15 - 7 =	15- 8 =	15 - 9 =	15 - 10 =
16 - 6 =	16 - 7 =	16 - 8 =	16 - 9 =	16 - 10 =
17 - 6 =	17 - 7 =	17 – 8 =	17- 9 =	17 - 10 =
18 - 6 =	18 - 7 =	18 - 8 =	18 - 9 =	18 - 10 =
19 - 6 =	19 - 7 =	19 - 8 =	19 - 9 =	19 - 10 =
20 - 6 =	20 - 7 =	20 - 8 =	20 - 9 =	20 - 10 =
21 - 6 =	21 - 7 =	21 - 8 =	21 - 9 =	21 - 10 =

22 - 6 =	22 - 7 =	22 - 8 =	22 - 9 =	22 - 10 =
23 - 6 =	23 - 7 =	23 - 8 =	23 - 9 =	23 - 10 =
24 - 6 =	24 - 7 =	24 - 8 =	24 - 9 =	24 - 10 =
25 - 6 =	25 - 7 =	25 - 8 =	25 - 9 =	25 - 10 =

11 - 0 =	12 - 0 =	13 - 0 =	14 - 0 =	15 - 0 =
11- 1 =	12 - 1 =	13 - 1 =	14 - 1 =	15 - 1 =
11- 2 =	12 - 2 =	13 - 2 =	14 - 2 =	15 - 2 =
11- 3 =	12 - 3 =	13 - 3 =	14 - 3 =	15 - 3 =
11- 4 =	12 - 4 =	13 - 4 =	14 - 4 =	15 - 4 =
11- 5 =	12 - 5 =	13 - 5 =	14 - 5 =	15 - 5 =
11- 6 =	12 - 6 =	13 - 6 =	14 - 6 =	15 - 6 =
11- 7 =	12 - 7 =	13 - 7 =	14 - 7 =	15 - 7 =
11- 8 =	12 - 8 =	13 - 8 =	14 - 8 =	15 - 8 =
11- 9 =	12 - 9 =	13 - 9 =	14 - 9 =	15 - 9 =
11- 10 =	12 - 10 =	13 - 10 =	14 - 10 =	15 - 10 =
11- 11 =	12 - 11 =	13 - 11 =	14 - 11 =	15 - 11 =
12 - 11 =	12 - 12 =	13 - 12 =	14 - 12 =	15 - 12 =
13 - 11 =	13 - 12 =	13 - 13 =	14 - 13 =	15 - 13 =
14 - 11 =	14 - 12 =	14 - 13 =	14 - 14 =	15 - 14 =
15 – 11 =	15 – 12 =	15 – 13 =	15 – 14 =	15 – 15 =
16 - 11 =	16 - 12 =	16 - 13 =	16 - 14 =	16 - 15 =
17 - 11 =	17- 12 =	17 - 13 =	17 - 14 =	17 - 15 =
18 - 11 =	18 - 12 =	18 - 13 =	18 - 14 =	18 - 15 =
19 - 11 =	19 - 12 =	19 - 13 =	19 - 14 =	19 - 15 =
20 - 11 =	20 - 12 =	20 - 13 =	20 - 14 =	20 - 15 =
21 – 11 =	21 – 12 =	21 – 13 =	21 – 14 =	21 – 15 =
22 - 11 =	22 - 12 =	22 - 13 =	22 - 14 =	22 - 15 =
23 - 11 =	23 - 12 =	23 - 13 =	23 - 14 =	23 -15 =
24 - 11 =	24 - 12 =	24 - 13 =	24 - 14 =	24 - 15 =

25 - 11 =	25 - 12 =	25 - 13 =	25 - 14 =	25 - 15 =

16 - 0 =	17 - 0 =	18 - 0 =	19 - 0 =	20 - 0 =
16 - 1 =	17- 1 =	18 - 1 =	19 - 1 =	20 - 1 =
16 - 2 =	17- 2 =	18 - 2 =	19 - 2 =	20 - 2 =
16 - 3 =	17- 3 =	18 - 3 =	19 - 3 =	20 - 3 =
16 - 4 =	17 - 4 =	18 - 4 =	19 - 4 =	20 - 4 =
16 - 5 =	17 - 5 =	18 - 5 =	19 - 5 =	20 - 5 =
16 - 6 =	17 - 6 =	18 - 6 =	19 - 6 =	20 - 6 =
16 - 7 =	17 - 7 =	18 - 7 =	19 - 7 =	20 - 7 =
16 - 8 =	17 - 8 =	18 - 8 =	19 - 8 =	20 - 8 =
16 - 9 =	17 - 9 =	18 - 9 =	19 - 9 =	20 - 9 =
16 -10 =	17 - 10 =	18 - 10 =	19 - 10 =	20 - 10 =
16 - 11 =	17 - 11 =	18 - 11 =	19 - 11 =	20 - 11 =
16 -12 =	17 - 12 =	18 - 12 =	19 - 12 =	20 - 12 =
16 - 13 =	17 - 13 =	18 - 13 =	19 - 13 =	20 - 13 =
16 - 14 =	17 -14 =	18 - 14 =	19 - 14 =	20 - 14 =
16 - 15 =	17- 15 =	18 - 15 =	19 - 15 =	20 - 15 =
16 - 16 =	17 - 16 =	18 - 16 =	19 - 16 =	20 - 16 =
17 - 16 =	17 - 17 =	18 - 17 =	19 - 17 =	20 - 17 =
18 - 16 =	18 - 17 =	18 - 18 =	19 - 18 =	20 - 18 =
19 - 16 =	19 - 17 =	18 - 18 =	19 - 19 =	20 - 19 =
20 - 16 =	20 - 17 =	20 - 18 =	20 - 19 =	20 - 20 =
21 - 16 =	21 - 17 =	21- 18 =	21- 19 =	21- 20 =
22 - 16 =	22 - 17 =	22 - 18 =	22- 19 =	22 - 20 =
23 - 16 =	23 - 17 =	23 - 18 =	23 - 19 =	23 - 20 =
24 - 16 =	24 - 17 =	24 - 18 =	24 - 19 =	24 - 20 =
25 - 16 =	25 -17 =	25 - 18 =	25 - 19 =	25 - 20 =

21 - 0 =	22 - 0 =	23 - 0 =	24 - 0 =	25 - 0 =

21 - 1 =	22 - 1 =	23 - 1 =	24 - 1 =	25 - 1 =
21 - 2 =	22 - 2 =	23 - 2 =	24 - 2 =	25 - 2 =
21 - 3 =	22 - 3 =	23 - 3 =	24 - 3 =	25 – 3 =
21 - 4 =	22 - 4 =	23 - 4 =	24 - 4 =	25 – 4 =
21 - 5 =	22 - 5 =	23 - 5 =	24 - 5 =	25 – 5 =
21 - 6 =	22 - 6 =	23 - 6 =	24 - 6 =	25 – 6 =
21 - 7 =	22 - 7 =	23 - 7 =	24 - 7 =	25 – 7 =
21 - 8 =	22 - 8 =	23 - 8 =	24 - 8 =	25 – 8 =
21 - 9 =	22 - 9 =	23 - 9 =	24 - 9 =	25 – 9 =
21 - 10 =	22 - 10 =	23 - 10 =	24 - 10 =	25 – 10 =
21 - 11 =	22 - 11 =	23 - 11 =	24 - 11 =	25 – 11 =
21 - 12 =	22 - 12 =	23 - 12 =	24 - 12 =	25 – 12 =
21- 13 =	22 - 13 =	23 - 13 =	24 - 13 =	25 – 13 =
21 - 14 =	22 - 14 =	23 - 14 =	24 - 14 =	25 – 14 =
21 - 15 =	22 - 15 =	23 -15 =	24 - 15 =	25 – 15 =
21 - 16 =	22 - 16 =	23 - 16 =	24 - 16 =	25 – 16 =
21 - 17 =	22 - 17 =	23 - 17 =	24 - 17 =	25 – 17 =
21 - 18 =	22 - 18 =	23 - 18 =	24 - 18 =	25 – 18 =
21 - 19 =	22 - 19 =	23 - 19 =	24 - 19 =	25 – 19 =
21 - 20 =	22 - 20 =	23 - 20 =	24 - 20 =	25 – 20 =
21 - 21 =	22 - 21 =	23 - 21 =	24 - 21 =	25 – 21 =
22 - 21 =	22 - 22 =	23 - 22 =	24 - 22 =	25 – 22 =
23 - 21 =	23- 22 =	23 - 23 =	24 - 23 =	25 – 23 =
24 - 21 =	24 - 22 =	24 - 23 =	24 - 24 =	25 – 24 =
25 - 21 =	25 - 22 =	25 - 23 =	24 - 23 =	25 – 25 =

MULTIPLYING NUMBERS

1 x 0 =	2 x 0 =	3 x 0 =	4 x 0 =	5 x 0 =
1 x 1 =	2 x 1 =	3 x 1 =	4 x 1 =	5 x 1 =
1 x 2 =	2 x 2 =	3 x 2 =	4 x 2 =	5 x 2 =
1 x 3 =	2 x 3 =	3 x 3 =	4 x 3 =	5 x 3 =
1 x 4 =	2 x 4 =	3 x 4 =	4 x 4 =	5 x 4 =
1 x 5 =	2 x 5 =	3 x 5 =	4 x 5 =	5 x 5 =
1 x 6 =	2 x 6 =	3 x 6 =	4 x 6 =	5 x 6 =
1 x 7 =	2 x 7 =	3 x 7 =	4 x 7 =	5 x 7 =
1 x 8 =	2 x 8 =	3 x 8 =	4 x 8 =	5 x 8 =
1 x 9 =	2 x 9 =	3 x 9 =	4 x 9 =	5 x 9 =
1 x 10 =	2 x 10 =	3 x 10 =	4 x 10 =	5 x 10 =
1 x 11 =	2 x 11 =	3 x 11 =	4 x 11 =	5 x 11 =
1 x 12 =	2 x 12 =	3 x 12 =	4 x 12 =	5 x 12 =
1 x 13 =	2 x 13 =	3 x 13 =	4 x 13 =	5 x 13 =
1 x 14 =	2 x 14 =	3 x 14 =	4 x 14 =	5 x 14 =
1 x 15 =	2 x 15 =	3 x 15 =	4 x 15 =	5 x 15 =
1 x 16 =	2 x 16 =	3 x 16 =	4 x 16 =	5 x 16 =
1 x 17 =	2 x 17 =	3 x 17 =	4 x 17 =	5 x 17 =
1 x 18 =	2 x 18 =	3 x 18 =	4 x 18 =	5 x 18 =
1 x 19 =	2 x 19 =	3 x 19 =	4 x 19 =	5 x 19 =

1 x 20 =	2 x 20 =	3 x 20 =	4 x 20 =	5 x 20 =
1 x 21 =	2 x 21 =	3 x 21 =	4 x 21 =	5 x 21 =
1 x 22 =	2 x 22 =	3 x 22 =	4 x 22 =	5 x 22 =
1 x 23 =	2 x 23 =	3 x 23 =	4 x 23 =	5 x 23 =
1 x 24 =	2 x 24 =	3 x 24 =	4 x 24 =	5 x 24 =
1 x 25 =	2 x 25 =	3 x 25 =	4 x 25 =	5 x 25 =

6 x 0 =	7 x 0 =	8 x 0 =	9 x 0 =
6 x 1 =	7 x 1 =	8 x 1 =	9 x 1 =
6 x 2 =	7 x 2 =	8 x 2 =	9 x 2 =
6 x 3 =	7 x 3 =	8 x 3 =	9 x 3 =
6 x 4 =	7 x 4 =	8 x 4 =	9 x 4 =
6 x 5 =	7 x 5 =	8 x 5 =	9 x 5 =
6 x 6 =	7 x 6 =	8 x 6 =	9 x 6 =
6 x 7 =	7 x 7 =	8 x 7 =	9 x 7 =
6 x 8 =	7 x 8 =	8 x 8 =	9 x 8 =
6 x 9 =	7 x 9 =	8 x 9 =	9 x 9 =
6 x 10 =	7 x 10 =	8 x 10 =	9 x 10 =
6 x 11 =	7 x 11 =	8 x 11 =	9 x 11 =
6 x 12 =	7 x 12 =	8 x 12 =	9 x 12 =
6 x 13 =	7 x 13 =	8 x 13 =	9 x 13 =
6 x 14 =	7 x 14 =	8 x 14 =	9 x 14 =
6 x 15 =	7 x 15 =	8 x 15 =	9 x 15 =
6 x 16 =	7 x 16 =	8 x 16 =	9 x 16 =
6 x 17 =	7 x 17 =	8 x 17 =	9 x 17 =
6 x 18 =	7 x 18 =	8 x 18 =	9 x 18 =
6 x 19 =	7 x 19 =	8 x 19 =	9 x 19 =
6 x 20 =	7 x 20 =	8 x 20 =	9 x 20 =
6 x 21 =	7 x 21 =	8 x 21 =	9 x 21 =
6 x 22 =	7 x 22 =	8 x 22 =	9 x 22 =

6 x 23 =	7 x 23 =	8 x 23 =	9 x 23 =
6 x 24 =	7 x 24 =	8 x 24 =	9 x 24 =
6 x 25 =	7 x 25 =	8 x 25 =	9 x 25 =

10 x 0 =	11 x 0 =	12 x 0 =	13 x 0 =
10 x 1 =	11 x 1 =	12 x 1 =	13 x 1 =
10 x 2 =	11 x 2 =	12 x 2 =	13 x 2 =
10 x 3 =	11 x 3 =	12 x 3 =	13 x 3 =
10 x 4 =	11 x 4 =	12 x 4 =	13 x 4 =
10 x 5 =	11 x 5 =	12 x 5 =	13 x 5 =
10 x 6 =	11 x 6 =	12 x 6 =	13 x 6 =
10 x 7 =	11 x 7 =	12 x 7 =	13 x 7 =
10 x 8 =	11 x 8 =	12 x 8 =	13 x 8 =
10 x 9 =	11 x 9 =	12 x 9 =	13 x 9 =
10 x 10 =	11 x 10 =	12 x 10 =	13 x 10 =
10 x 11 =	11 x 11 =	12 x 11 =	13 x 11 =
10 x 12 =	11 x 12 =	12 x 12 =	13 x 12 =
10 x 13 =	11 x 13 =	12 x 13 =	13 x 13 =
10 x 14 =	11 x 14 =	12 x 14 =	13 x 14 =
10 x 15 =	11 x 15 =	12 x 15 =	13 x 15 =
10 x 16 =	11 x 16 =	12 x 16 =	13 x 16 =
10 x 17 =	11 x 17 =	12 x 17 =	13 x 17 =
10 x 18 =	11 x 18 =	12 x 18 =	13 x 18 =
10 x 19 =	11 x 19 =	12 x 19 =	13 x 19 =
10 x 20 =	11 x 20 =	12 x 20 =	13 x 20 =
10 x 21 =	11 x 21 =	12 x 21 =	13 x 21 =
10 x 22 =	11 x 22 =	12 x 22 =	13 x 22 =
10 x 23 =	11 x 23 =	12 x 23 =	13 x 23 =
10 x 24 =	11 x 24 =	12 x 24 =	13 x 24 =
10 x 25 =	11 x 25 =	12 x 25 =	13 x 25 =

14 x 0 =	15 x 0 =	16 x 0 =	17 x 0 =
14 x 1 =	15 x 1 =	16 x 1 =	17 x 1 =
14 x 2 =	15 x 2 =	16 x 2 =	17 x 2 =
14 x 3 =	15 x 3 =	16 x 3 =	17 x 3 =
14 x 4 =	15 x 4 =	16 x 4 =	17 x 4 =
14 x 5 =	15 x 5 =	16 x 5 =	17 x 5 =
14 x 6 =	15 x 6 =	16 x 6 =	17 x 6 =
14 x 7 =	15 x 7 =	16 x 7 =	17 x 7 =
14 x 8 =	15 x 8 =	16 x 8 =	17 x 8 =
14 x 9 =	15 x 9 =	16 x 9 =	17 x 9 =
14 x 10 =	15 x 10 =	16 x 10 =	17 x 10 =
14 x 11 =	15 x 11 =	16 x 11 =	17 x 11 =
14 x 12 =	15 x 12 =	16 x 12 =	17 x 12 =
14 x 13 =	15 x 13 =	16 x 13 =	17 x 13 =
14 x 14 =	15 x 14 =	16 x 14 =	17 x 14 =
14 x 15 =	15 x 15 =	16 x 15 =	17 x 15 =
14 x 16 =	15 x 16 =	16 x 16 =	17 x 16 =
14 x 17 =	15 x 17 =	16 x 17 =	17 x 17 =
14 x 18 =	15 x 18 =	16 x 18 =	17 x 18 =
14 x 19 =	15 x 19 =	16 x 19 =	17 x 19 =
14 x 20 =	15 x 20 =	16 x 20 =	17 x 20 =
14 x 21 =	15 x 21 =	16 x 21 =	17 x 21 =
14 x 22 =	15 x 22 =	16 x 22 =	17 x 22 =
14 x 23 =	15 x 23 =	16 x 23 =	17 x 23 =
14 x 24 =	15 x 24 =	16 x 24 =	17 x 24 =
14 x 25 =	15 x 25 =	16 x 25 =	17 x 25 =

18 x 0 =	19 x 0 =	20 x 0 =	21 x 0 =
18 x 1 =	19 x 1 =	20 x 1 =	21 x 1 =

18 x 2 =	19 x 2 =	20 x 2 =	21 x 2 =
18 x 3 =	19 x 3 =	20 x 3 =	21 x 3 =
18 x 4 =	19 x 4 =	20 x 4 =	21 x 4 =
18 x 5 =	19 x 5 =	20 x 5 =	21 x 5 =
18 x 6 =	19 x 6 =	20 x 6 =	21 x 6 =
18 x 7 =	19 x 7 =	20 x 7 =	21 x 7 =
18 x 8 =	19 x 8 =	20 x 8 =	21 x 8 =
18 x 9 =	19 x 9 =	20 x 9 =	21 x 9 =
18 x 10 =	19 x 10 =	20 x 10 =	21 x 10 =
18 x 11 =	19 x 11 =	20 x 11 =	21 x 11 =
18 x 12 =	19 x 12 =	20 x 12 =	21 x 12 =
18 x 13 =	19 x 13 =	20 x 13 =	21 x 13 =
18 x 14 =	19 x 14 =	20 x 14 =	21 x 14 =
18 x 15 =	19 x 15 =	20 x 15 =	21 x 15 =
18 x 16 =	19 x 16 =	20 x 16 =	21 x 16 =
18 x 17 =	19 x 17 =	20 x 17 =	21 x 17 =
18 x 18 =	19 x 18 =	20 x 18 =	21 x 18 =
18 x 19 =	19 x 19 =	20 x 19 =	21 x 19 =
18 x 20 =	19 x 20 =	20 x 20 =	21 x 20 =
18 x 21 =	19 x 21 =	20 x 21 =	21 x 21 =
18 x 22 =	19 x 22 =	20 x 22 =	21 x 22 =
18 x 23 =	19 x 23 =	20 x 23 =	21 x 23 =
18 x 24 =	19 x 24 =	20 x 24 =	21 x 24 =
18 x 25 =	19 x 25 =	20 x 25 =	21 x 25 =

22 x 0 =	23 x 0 =	24 x 0 =	25 x 0 =
22 x 1 =	23 x 1 =	24 x 1 =	25 x 1 =
22 x 2 =	23 x 2 =	24 x 2 =	25 x 2 =
22 x 3 =	23 x 3 =	24 x 3 =	25 x 3 =
22 x 4 =	23 x 4 =	24 x 4 =	25 x 4 =

22 x 5 =	23 x 6 =	24 x 6 =	25 x 6 =
22 x 7 =	23 x 7 =	24 x 7 =	25 x 7 =
22 x 8 =	23 x 8 =	24 x 8 =	25 x 8 =
22 x 9 =	23 x 9 =	24 x 9 =	25 x 9 =
22 x 10 =	23 x 10 =	24 x 10 =	25 x 10 =
22 x 11 =	23 x 11 =	24 x 11 =	25 x 11 =
22 x 12 =	23 x 12 =	24 x 12 =	25 x 12 =
22 x 13 =	23 x 13 =	24 x 13 =	25 x 13 =
22 x 14 =	23 x 14 =	24 x 14 =	25 x 14 =
22 x 15 =	23 x 15 =	24 x 15 =	25 x 15 =
22 x 16 =	23 x 16 =	24 x 16 =	25 x 16 =
22 x 17 =	23 x 17 =	24 x 17 =	25 x 17 =
22 x 18 =	23 x 18 =	24 x 18 =	25 x 18 =
22 x 19 =	23 x 19 =	24 x 19 =	25 x 19 =
22 x 20 =	23 x 20 =	24 x 20 =	25 x 20 =
22 x 21 =	23 x 21 =	24 x 21 =	25 x 21 =
22 x 22 =	23 x 22 =	24 x 22 =	25 x 22 =
22 x 23 =	23 x 23 =	24 x 23 =	25 x 23 =
22 x 24 =	23 x 24 =	24 x 24 =	25 x 24 =
22 x 25 =	23 x 25 =	24 x 25 =	25 x 25 =

DIVIDING NUMBERS

1 ÷ 0 =	2 ÷ 0 =	3 ÷ 0 =	4 ÷ 0 =	5 ÷ 0 =
1 ÷ 1 =	2 ÷ 2 =	3 ÷ 3 =	4 ÷ 4 =	5 ÷ 5 =
2 ÷ 1 =	4 ÷ 2 =	6 ÷ 3 =	8 ÷ 4 =	10 ÷ 5 =
3 ÷ 1 =	6 ÷ 2 =	9 ÷ 3 =	12 ÷ 4 =	15 ÷ 5 =
4 ÷ 1 =	8 ÷ 2 =	12 ÷ 3 =	16 ÷ 4 =	20 ÷ 5 =
5 ÷ 1 =	10 ÷ 2 =	15 ÷ 3 =	20 ÷ 4 =	25 ÷ 5 =
6 ÷ 1 =	12 ÷ 2 =	18 ÷ 3 =	24 ÷ 4 =	30 ÷ 5 =
7 ÷ 1 =	14 ÷ 2 =	21 ÷ 3 =	28 ÷ 4 =	35 ÷ 5 =
8 ÷ 1 =	16 ÷ 2 =	24 ÷ 3 =	32 ÷ 4 =	40 ÷ 5 =
9 ÷ 1 =	18 ÷ 2 =	27 ÷ 3 =	36 ÷ 4 =	45 ÷ 5 =
10 ÷ 1 =	20 ÷ 2 =	30 ÷ 3 =	40 ÷ 4 =	50 ÷ 5 =
11 ÷ 1 =	22 ÷ 2 =	33 ÷ 3 =	44 ÷ 4 =	55 ÷5 =
12 ÷ 1 =	24 ÷ 2 =	36 ÷ 3 =	48 ÷ 4 =	60 ÷ 5 =
13 ÷ 1 =	26 ÷ 2 =	39 ÷ 3 =	52 ÷ 4 =	65 ÷ 5 =
14 ÷ 1 =	28 ÷ 2 =	42 ÷ 3 =	56 ÷ 4 =	70 ÷ 5 =
15 ÷ 1 =	30 ÷ 2 =	45 ÷ 3 =	60 ÷ 4 =	75 ÷ 5 =
16 ÷ 1 =	32 ÷ 2 =	48 ÷ 3 =	64 ÷ 4 =	80 ÷ 5 =
17 ÷ 1 =	34 ÷ 2 =	51 ÷ 3 =	68 ÷ 4 =	85 ÷ 5 =
18 ÷ 1 =	36 ÷ 2 =	54 ÷ 3 =	72 ÷ 4 =	90 ÷ 5 =
19 ÷ 1 =	38 ÷ 2 =	57 ÷ 3 =	76 ÷ 4 =	95 ÷ 5 =

$20 \div 1 =$	$40 \div 2 =$	$60 \div 3 =$	$80 \div 4 =$	$100 \div 5 =$
$21 \div 1 =$	$42 \div 2 =$	$63 \div 3 =$	$84 \div 4 =$	$105 \div 5 =$
$22 \div 1 =$	$44 \div 2 =$	$66 \div 3 =$	$88 \div 4 =$	$110 \div 5 =$
$23 \div 1 =$	$46 \div 2 =$	$69 \div 3 =$	$92 \div 4 =$	$115 \div 5 =$
$24 \div 1 =$	$48 \div 2 =$	$72 \div 3 =$	$96 \div 4 =$	$120 \div 5 =$
$25 \div 1 =$	$50 \div 2 =$	$75 \div 3 =$	$100 \div 4 =$	$125 \div 5 =$

$6 \div 0 =$	$7 \div 0 =$	$8 \div 0 =$	$9 \div 0 =$
$6 \div 6 =$	$7 \div 7 =$	$8 \div 8 =$	$9 \div 9 =$
$12 \div 6 =$	$14 \div 7 =$	$16 \div 8 =$	$18 \div 9 =$
$18 \div 6 =$	$21 \div 7 =$	$24 \div 8 =$	$27 \div 9 =$
$24 \div 6 =$	$28 \div 7 =$	$32 \div 8 =$	$36 \div 9 =$
$30 \div 6 =$	$35 \div 7 =$	$40 \div 8 =$	$45 \div 9 =$
$36 \div 6 =$	$42 \div 7 =$	$48 \div 8 =$	$54 \div 9 =$
$42 \div 6 =$	$49 \div 7 =$	$56 \div 8 =$	$63 \div 9 =$
$48 \div 6 =$	$56 \div 7 =$	$64 \div 8 =$	$72 \div 9 =$
$54 \div 6 =$	$63 \div 7 =$	$72 \div 8 =$	$81 \div 9 =$
$60 \div 6 =$	$70 \div 7 =$	$80 \div 8 =$	$90 \div 9 =$
$66 \div 6 =$	$77 \div 7 =$	$88 \div 8 =$	$99 \div 9 =$
$72 \div 6 =$	$84 \div 7 =$	$96 \div 8 =$	$108 \div 9 =$
$78 \div 6 =$	$91 \div 7 =$	$104 \div 8 =$	$117 \div 9 =$
$84 \div 6 =$	$98 \div 7 =$	$112 \div 8 =$	$126 \div 9 =$
$90 \div 6 =$	$105 \div 7 =$	$120 \div 8 =$	$135 \div 9 =$
$96 \div 6 =$	$112 \div 7 =$	$128 \div 8 =$	$144 \div 9 =$
$102 \div 6 =$	$119 \div 7 =$	$136 \div 8 =$	$153 \div 9 =$
$108 \div 6 =$	$126 \div 7 =$	$144 \div 8 =$	$162 \div 9 =$
$114 \div 6 =$	$133 \div 7 =$	$152 \div 8 =$	$171 \div 9 =$
$120 \div 6 =$	$140 \div 7 =$	$160 \div 8 =$	$180 \div 9 =$
$126 \div 6 =$	$147 \div 7 =$	$168 \div 8 =$	$189 \div 9 =$
$132 \div 6 =$	$154 \div 7 =$	$176 \div 8 =$	$198 \div 9 =$

138 ÷ 6 =	161 ÷ 7 =	184 ÷ 8 =	207 ÷ 9 =
144 ÷ 6 =	168 ÷ 7 =	192 ÷ 8 =	216 ÷ 9 =
150 ÷ 6 =	175 ÷7 =	200 ÷ 8 =	225 ÷ 9 =

10 ÷ 0 =	11 ÷ 0 =	12 ÷ 0 =	13 ÷ 0 =
10 ÷ 10 =	11 ÷ 11 =	12 ÷ 12 =	13 ÷ 13 =
20 ÷ 10 =	22 ÷ 11 =	24 ÷ 12 =	26 ÷ 13 =
30 ÷ 10 =	33 ÷ 11 =	36 ÷ 12 =	39 ÷ 13 =
40 ÷ 10 =	44 ÷ 11 =	48 ÷ 12 =	52 ÷ 13 =
50 ÷ 10 =	55 ÷ 11 =	60 ÷ 12 =	65 ÷ 13 =
60 ÷ 10 =	66 ÷ 11 =	72 ÷ 12 =	78 ÷ 13 =
70 ÷ 10 =	77 ÷ 11 =	84 ÷ 12 =	91 ÷ 13 =
80 ÷ 10 =	88 ÷ 11 =	96 ÷ 12 =	104 ÷ 13 =
90 ÷ 10 =	99 ÷ 11 =	108 ÷ 12 =	117 ÷ 13 =
100 ÷ 10 =	110 ÷ 11 =	120 ÷ 12 =	130 ÷ 13 =
110 ÷ 10 =	121 ÷ 11 =	132 ÷ 12 =	143 ÷ 13 =
120 ÷ 10 =	132 ÷ 11 =	144 ÷ 12 =	156 ÷ 13 =
130 ÷ 10 =	143 ÷ 11 =	156 ÷ 12 =	169 ÷ 13 =
140 ÷ 10 =	154 ÷ 11 =	168 ÷ 12 =	182 ÷ 13 =
150 ÷ 10 =	165 ÷ 11 =	180 ÷ 12 =	195 ÷ 13 =
160 ÷ 10 =	176 ÷ 11 =	192 ÷ 12 =	208 ÷ 13 =
170 ÷ 10 =	187 ÷ 11 =	204 ÷ 12 =	221 ÷ 13 =
180 ÷ 10 =	198 ÷ 11 =	216 ÷ 12 =	234 ÷ 13 =
190 ÷ 10 =	209 ÷ 11 =	228 ÷ 12 =	247 ÷ 13 =
200 ÷ 10 =	220 ÷ 11 =	240 ÷ 12 =	260 ÷ 13 =
210 ÷ 10 =	231 ÷ 11 =	252 ÷ 12 =	273 ÷ 13 =
220 ÷ 10 =	242 ÷ 11 =	264 ÷ 12 =	286 ÷ 13 =
230 ÷ 10 =	253 ÷ 11 =	276 ÷ 12 =	299 ÷ 13 =
240 ÷ 10 =	264 ÷ 11 =	288 ÷ 12 =	312 ÷ 13 =
250 ÷ 10 =	275 ÷ 11 =	300 ÷ 12 =	325 ÷ 13 =

14 ÷ 0 =	15 ÷ 0 =	16 ÷ 0 =	17 ÷ 0 =
14 ÷ 14 =	15 ÷ 15 =	16 ÷ 16 =	17 ÷ 17 =
28 ÷ 14 =	30 ÷ 15 =	32 ÷ 16 =	34 ÷ 17 =
42 ÷ 14 =	45 ÷ 15 =	48 ÷ 16 =	51 ÷ 17 =
56 ÷ 14 =	60 ÷ 15 =	64 ÷ 16 =	68 ÷ 17 =
70 ÷ 14 =	75 ÷ 15 =	80 ÷ 16 =	85 ÷ 17 =
84 ÷ 14 =	90 ÷ 15 =	96 ÷ 16 =	102 ÷ 17 =
98 ÷ 14 =	105 ÷ 15 =	112 ÷ 16 =	119 ÷ 17 =
112 ÷ 14 =	120 ÷ 15 =	128 ÷ 16 =	136 ÷ 17 =
126 ÷ 14 =	135 ÷ 15 =	144 ÷ 16 =	153 ÷ 17 =
140 ÷ 14 =	150 ÷ 15 =	160 ÷ 16 =	170 ÷ 17 =
154 ÷ 14 =	165 ÷ 15 =	176 ÷ 16 =	187 ÷ 17 =
168 ÷ 14 =	180 ÷ 15 =	192 ÷ 16 =	204 ÷ 17 =
182 ÷ 14 =	195 ÷ 15 =	208 ÷ 16 =	221 ÷ 17 =
196 ÷ 14 =	210 ÷ 15 =	224 ÷ 16 =	238 ÷ 17 =
210 ÷ 14 =	225 ÷ 15 =	240 ÷ 16 =	255 ÷ 17 =
224 ÷ 14 =	240 ÷ 15 =	256 ÷ 16 =	272 ÷ 17 =
238 ÷ 14 =	255 ÷ 15 =	272 ÷ 16 =	289 ÷ 17 =
252 ÷ 14 =	270 ÷ 15 =	288 ÷ 16 =	306 ÷ 17 =
266 ÷ 14 =	285 ÷ 15 =	304 ÷ 16 =	323 ÷ 17 =
280 ÷ 14 =	300 ÷ 15 =	320 ÷ 16 =	340 ÷ 17 =
294 ÷ 14 =	315 ÷ 15 =	336 ÷ 16 =	357 ÷ 17 =
308 ÷ 14 =	330 ÷ 15 =	352 ÷ 16 =	374 ÷ 17 =
322 ÷ 14 =	345 ÷ 15 =	368 ÷ 16 =	391 ÷ 17 =
336 ÷ 14 =	360 ÷ 15 =	384 ÷ 16 =	408 ÷ 17 =
350 ÷ 14 =	375 ÷ 15 =	400 ÷ 16 =	425 ÷ 17 =

18 ÷ 0 =	19 ÷ 0 =	20 ÷ 0 =	21 ÷ 0 =
18 ÷ 18 =	19 ÷ 19 =	20 ÷ 20 =	21 ÷ 21 =

36 ÷ 18 =	38 ÷ 19 =	40 ÷ 20 =	42 ÷ 21 =
54 ÷ 18 =	57 ÷ 19 =	60 ÷ 20 =	63 ÷ 21 =
72 ÷ 18 =	76 ÷ 19 =	80 ÷ 20 =	84 ÷ 21 =
90 ÷ 18 =	95 ÷ 19 =	100 ÷ 20 =	105 ÷ 21 =
108 ÷ 18 =	114 ÷ 19 =	120 ÷ 20 =	126 ÷ 21 =
126 ÷ 18 =	133 ÷ 19 =	140 ÷ 20 =	147 ÷ 21 =
144 ÷ 18 =	152 ÷ 19 =	160 ÷ 20 =	168 ÷ 21 =
162 ÷ 18 =	171 ÷ 19 =	180 ÷ 20 =	189 ÷ 21 =
180 ÷ 18 =	190 ÷ 19 =	200 ÷ 20 =	210 ÷ 21 =
198 ÷ 18 =	209 ÷ 19 =	220 ÷ 20 =	231 ÷ 21 =
216 ÷ 18 =	228 ÷ 19 =	240 ÷ 20 =	252 ÷ 21 =
234 ÷ 18 =	247 ÷ 19 =	260 ÷ 20 =	273 ÷ 21 =
252 ÷ 18 =	266 ÷ 19 =	280 ÷ 20 =	294 ÷ 21 =
270 ÷ 18 =	285 ÷ 19 =	300 ÷ 20 =	315 ÷ 21 =
288 ÷ 18 =	304 ÷ 19 =	320 ÷ 20 =	336 ÷ 21 =
306 ÷ 18 =	323 ÷ 19 =	340 ÷ 20 =	357 ÷ 21 =
324 ÷ 18 =	342 ÷ 19 =	360 ÷ 20 =	378 ÷ 21 =
342 ÷ 18 =	361 ÷ 19 =	380 ÷ 20 =	399 ÷ 21 =
360 ÷ 18 =	380 ÷ 19 =	400 ÷ 20 =	420 ÷ 21 =
378 ÷ 18 =	399 ÷ 19 =	420 ÷ 20 =	441 ÷ 21 =
396 ÷ 18 =	418 ÷ 19 =	440 ÷ 20 =	462 ÷ 21 =
414 ÷ 18 =	437 ÷ 19 =	460 ÷ 20 =	483 ÷ 21 =
432 ÷ 18 =	456 ÷ 19 =	480 ÷ 20 =	504 ÷ 21 =
450 ÷ 18 =	475 ÷ 19 =	500 ÷ 20 =	525 ÷ 21 =

22 ÷ 0 =	23 ÷ 0 =	24 ÷ 0 =	25 ÷ 0 =
22 ÷ 22 =	23 ÷ 23 =	24 ÷ 24 =	25 ÷ 25 =
44 ÷ 22 =	46 ÷ 23 =	48 ÷ 24 =	50 ÷ 25 =
66 ÷ 22 =	69 ÷ 23 =	72 ÷ 24 =	75 ÷ 25 =
88 ÷ 22 =	92 ÷ 23 =	96 ÷ 24 =	100 ÷ 25 =

110 ÷ 22 =	115 ÷ 23 =	120 ÷ 24 =	125 ÷ 25 =
132 ÷ 22 =	138 ÷ 23 =	144 ÷ 24 =	150 ÷ 25 =
154 ÷ 22 =	161 ÷ 23 =	168 ÷ 24 =	175 ÷ 25 =
176 ÷ 22 =	184 ÷ 23 =	192 ÷ 24 =	200 ÷ 25 =
198 ÷ 22 =	207 ÷ 23 =	216 ÷ 24 =	225 ÷ 25 =
220 ÷ 22 =	230 ÷ 23 =	240 ÷ 24 =	250 ÷ 25 =
242 ÷ 22 =	253 ÷ 23 =	264 ÷ 24 =	275 ÷ 25 =
264 ÷ 22 =	276 ÷ 23 =	288 ÷ 24 =	300 ÷ 25 =
286 ÷ 22 =	299 ÷ 23 =	312 ÷ 24 =	325 ÷ 25 =
308 ÷ 22 =	322 ÷ 23 =	336 ÷ 24 =	350 ÷ 25 =
330 ÷ 22 =	345 ÷ 23 =	360 ÷ 24 =	375 ÷ 25 =
352 ÷ 22 =	368 ÷ 23 =	384 ÷ 24 =	400 ÷ 25 =
374 ÷ 22 =	391 ÷ 23 =	408 ÷ 24 =	425 ÷ 25 =
396 ÷ 22 =	414 ÷ 23 =	432 ÷ 24 =	450 ÷ 25 =
418 ÷ 22 =	437 ÷ 23 =	456 ÷ 24 =	475 ÷ 25 =
440 ÷ 22 =	460 ÷ 23 =	480 ÷ 24 =	500 ÷ 25 =
462 ÷ 22 =	483 ÷ 23 =	504 ÷ 24 =	525 ÷ 25 =
484 ÷ 22 =	506 ÷ 23 =	528 ÷ 24 =	550 ÷ 25 =
506 ÷ 22 =	529 ÷ 23 =	552 ÷ 24 =	575 ÷ 25 =
528 ÷ 22 =	552 ÷ 23 =	576 ÷ 24 =	600 ÷ 25 =
550 ÷ 22 =	575 ÷ 23 =	600 ÷ 24 =	625 ÷ 25 =

DECIMALS

0 ÷ 1 =	0 ÷ 2 =	0 ÷ 3 =	0 ÷ 4 =	0 ÷ 5 =
1 ÷ 1 =	1 ÷ 2 =	1 ÷ 3 =	1÷ 4 =	1 ÷ 5 =
2 ÷ 1 =	2 ÷ 2 =	2 ÷ 3 =	2 ÷ 4 =	2 ÷ 5 =
3 ÷ 1 =	3 ÷ 2 =	3 ÷ 3 =	3 ÷ 4 =	3 ÷ 5 =
4 ÷ 1 =	4 ÷ 2 =	4 ÷ 3 =	4 ÷ 4 =	4 ÷ 5 =
5 ÷ 1 =	5 ÷ 2 =	5 ÷ 3 =	5 ÷ 4 =	5 ÷ 5 =
6 ÷ 1 =	6 ÷ 2 =	6 ÷ 3 =	6 ÷ 4 =	6 ÷ 5 =
7 ÷ 1 =	7 ÷ 2 =	7 ÷ 3 =	7 ÷ 4 =	7 ÷ 5 =
8 ÷ 1 =	8 ÷ 2 =	8 ÷ 3 =	8 ÷ 4 =	8 ÷ 5 =
9 ÷ 1 =	9 ÷ 2 =	9 ÷ 3 =	9 ÷ 4 =	9 ÷ 5 =
10 ÷ 1 =	10 ÷ 2 =	10 ÷ 3 =	10 ÷ 4 =	10 ÷ 5 =
11 ÷ 1 =	11 ÷ 2 =	11 ÷ 3 =	11 ÷ 4 =	11 ÷ 5 =
12 ÷ 1 =	12 ÷ 2 =	12 ÷ 3 =	12 ÷ 4 =	12 ÷ 5 =
13 ÷ 1 =	13 ÷ 2 =	13 ÷ 3 =	13 ÷ 4 =	13 ÷ 5 =
14 ÷ 1 =	14 ÷ 2 =	14 ÷ 3 =	14 ÷ 4 =	14 ÷ 5 =
15 ÷ 1 =	15 ÷ 2 =	15 ÷ 3 =	15 ÷ 4 =	15 ÷ 5 =
16 ÷ 1 =	16 ÷ 2 =	16 ÷ 3 =	16 ÷ 4 =	16 ÷ 5 =
17 ÷ 1 =	17 ÷ 2 =	17 ÷ 3 =	17 ÷ 4 =	17 ÷ 5 =
18 ÷ 1 =	18 ÷ 2 =	18 ÷ 3 =	18 ÷ 4 =	18 ÷ 5 =
19 ÷ 1 =	19 ÷ 2 =	19 ÷ 3 =	19 ÷ 4 =	19 ÷ 5 =

20 ÷ 1 =	20 ÷ 2 =	20 ÷ 3 =	20 ÷ 4 =	20 ÷ 5 =
21 ÷ 1 =	21 ÷ 2 =	21 ÷ 3 =	21 ÷ 4 =	21 ÷ 5 =
22 ÷ 1 =	22 ÷ 2 =	22 ÷ 3 =	22 ÷ 4 =	22 ÷ 5 =
23 ÷ 1 =	23 ÷ 2 =	23 ÷ 3 =	23 ÷ 4 =	23 ÷ 5 =
24 ÷ 1 =	24 ÷ 2 =	24 ÷ 3 =	24 ÷ 4 =	24 ÷ 5 =
25 ÷ 1 =	25 ÷ 2 =	25 ÷ 3 =	25 ÷ 4 =	25 ÷ 5 =

0 ÷ 6 =	0 ÷ 7 =	0 ÷ 8 =	0 ÷ 9 =
1÷ 6 =	1 ÷ 7 =	1 ÷ 8 =	1 ÷ 9 =
2 ÷ 6 =	2 ÷ 7 =	2 ÷ 8 =	2 ÷ 9 =
3 ÷ 6 =	3 ÷ 7 =	3 ÷ 8 =	3 ÷ 9 =
4 ÷ 6 =	4 ÷ 7 =	4 ÷ 8 =	4 ÷ 9 =
5 ÷ 6 =	5 ÷ 7 =	5 ÷ 8 =	5 ÷ 9 =
6 ÷ 6 =	6 ÷ 7 =	6 ÷ 8 =	6 ÷ 9 =
7 ÷ 6 =	7 ÷ 7 =	7 ÷ 8 =	7 ÷ 9 =
8 ÷ 6 =	8 ÷ 7 =	8 ÷ 8 =	8 ÷ 9 =
9 ÷ 6 =	9 ÷ 7 =	9 ÷ 8 =	9 ÷ 9 =
10 ÷ 6 =	10 ÷ 7 =	10 ÷ 8 =	10 ÷ 9 =
11 ÷ 6 =	11 ÷ 7 =	11 ÷ 8 =	11 ÷ 9 =
12 ÷ 6 =	12 ÷ 7 =	12 ÷ 8 =	12 ÷ 9 =
13 ÷ 6 =	13 ÷ 7 =	13 ÷ 8 =	13 ÷ 9 =
14 ÷ 6 =	14 ÷ 7 =	14 ÷ 8 =	14 ÷ 9 =
15 ÷ 6 =	15 ÷ 7 =	15 ÷ 8 =	15 ÷ 9 =
16 ÷ 6 =	16 ÷ 7 =	16 ÷ 8 =	16 ÷ 9 =
17 ÷ 6 =	17 ÷ 7 =	17 ÷ 8 =	17 ÷ 9 =
18 ÷ 6 =	18 ÷ 7 =	18 ÷ 8 =	18 ÷ 9 =
19 ÷ 6 =	19 ÷ 7 =	19 ÷ 8 =	19 ÷ 9 =
20 ÷ 6 =	20 ÷ 7 =	20 ÷ 8 =	20 ÷ 9 =
21 ÷ 6 =	21 ÷ 7 =	21 ÷ 8 =	21 ÷ 9 =
22 ÷ 6 =	22 ÷ 7 =	22 ÷ 8 =	22 ÷ 9 =

23 ÷ 6 =	23 ÷ 7 =	23 ÷ 8 =	23 ÷ 9 =
24 ÷ 6 =	24 ÷ 7 =	24 ÷ 8 =	24 ÷ 9 =
25 ÷ 6 =	25 ÷ 7 =	25 ÷ 8 =	25 ÷ 9 =

0 ÷ 10 =	0 ÷ 11 =	0 ÷ 12 =	0 ÷ 13 =
1 ÷ 10 =	1 ÷ 11 =	1 ÷ 12 =	1 ÷ 13 =
2 ÷ 10 =	2 ÷ 11 =	2 ÷ 12 =	2 ÷ 13 =
3 ÷ 10 =	3 ÷ 11 =	3 ÷ 12 =	3 ÷ 13 =
4 ÷ 10 =	4 ÷ 11 =	4 ÷ 12 =	4 ÷ 13 =
5 ÷ 10 =	5 ÷ 11 =	5 ÷ 12 =	5 ÷ 13 =
6 ÷ 10 =	6 ÷ 11 =	6 ÷ 12 =	6 ÷ 13 =
7 ÷ 10 =	7 ÷ 11 =	7 ÷ 12 =	7 ÷ 13 =
8 ÷ 10 =	8 ÷ 11 =	8 ÷ 12 =	8 ÷ 13 =
9 ÷ 10 =	9 ÷ 11 =	9 ÷ 12 =	9 ÷ 13 =
10 ÷ 10 =	10 ÷ 11 =	10 ÷ 12 =	10 ÷ 13 =
11 ÷ 10 =	11 ÷11 =	11 ÷ 12 =	11 ÷ 13 =
12 ÷ 10 =	12 ÷ 11 =	12 ÷ 12 =	12 ÷ 13 =
13 ÷ 10 =	13 ÷ 11 =	13 ÷ 12 =	13 ÷ 13 =
14 ÷ 10 =	14 ÷ 11 =	14 ÷ 12 =	14 ÷ 13 =
15 ÷ 10 =	15 ÷ 11 =	15 ÷ 12 =	15 ÷ 13 =
16 ÷ 10 =	16 ÷ 11 =	16 ÷ 12 =	16 ÷ 13 =
17 ÷ 10 =	17 ÷ 11 =	17 ÷ 12 =	17 ÷ 13 =
18 ÷ 10 =	18 ÷ 11 =	18 ÷ 12 =	18 ÷ 13 =
19 ÷ 10 =	19 ÷ 11 =	19 ÷ 12 =	19 ÷ 13 =
20 ÷ 10 =	20 ÷ 11 =	20 ÷ 12 =	20 ÷ 13 =
21 ÷ 10 =	21 ÷ 11 =	21 ÷ 12 =	21 ÷ 13 =
22 ÷ 10 =	22 ÷ 11 =	22 ÷ 12 =	22 ÷ 13 =
23 ÷ 10 =	23 ÷ 11 =	23 ÷ 12 =	23 ÷ 13 =
24 ÷ 10 =	24 ÷ 11 =	24 ÷ 12 =	24 ÷ 13 =
25 ÷ 10 =	25 ÷ 11 =	25 ÷ 12 =	25 ÷ 13 =

0 ÷ 14 =	0 ÷ 15 =	0 ÷ 16 =	0 ÷ 17 =
1 ÷ 14 =	1 ÷ 15 =	1 ÷ 16 =	1 ÷ 17 =
2 ÷ 14 =	2 ÷ 15 =	2 ÷ 16 =	2 ÷ 17 =
3 ÷ 14 =	3 ÷ 15 =	3 ÷ 16 =	3 ÷ 17 =
4 ÷ 14 =	4 ÷ 15 =	4 ÷ 16 =	4 ÷ 17 =
5 ÷ 14 =	5 ÷ 15 =	5 ÷ 16 =	5 ÷ 17 =
6 ÷ 14 =	6 ÷ 15 =	6 ÷ 16 =	6 ÷ 17 =
7 ÷ 14 =	7 ÷ 15 =	7 ÷ 16 =	7 ÷ 17 =
8 ÷ 14 =	8 ÷ 15 =	8 ÷ 16 =	8 ÷ 17 =
9 ÷ 14 =	9 ÷ 15 =	9 ÷ 16 =	9 ÷ 17 =
10 ÷ 14 =	10 ÷ 15 =	10 ÷ 16 =	10 ÷ 17 =
11 ÷ 14 =	11 ÷ 15 =	11 ÷ 16 =	11 ÷ 17 =
12 ÷ 14 =	12 ÷ 15 =	12 ÷ 16 =	12 ÷ 17 =
13 ÷ 14 =	13 ÷ 15 =	13 ÷ 16 =	13 ÷ 17 =
14 ÷ 14 =	14 ÷ 15 =	14 ÷ 16 =	14 ÷ 17 =
15 ÷ 14 =	15 ÷ 15 =	15 ÷ 16 =	15 ÷ 17 =
16 ÷ 14 =	16 ÷ 15 =	16 ÷ 16 =	16 ÷ 17 =
17 ÷ 14 =	17 ÷ 15 =	17 ÷ 16 =	17 ÷ 17 =
18 ÷ 14 =	18 ÷ 15 =	18 ÷ 16 =	18 ÷ 17 =
19 ÷ 14 =	19 ÷ 15 =	19 ÷ 16 =	19 ÷ 17 =
20 ÷ 14 =	20 ÷ 15 =	20 ÷ 16 =	20 ÷ 17 =
21 ÷ 14 =	21 ÷ 15 =	21 ÷ 16 =	21 ÷ 17 =
22 ÷ 14 =	22 ÷ 15 =	22 ÷ 16 =	22 ÷ 17 =
23 ÷ 14 =	23 ÷ 15 =	23 ÷ 16 =	23 ÷ 17 =
24 ÷ 14 =	24 ÷ 15 =	24 ÷ 16 =	24 ÷ 17 =
25 ÷ 14 =	25 ÷ 15 =	25 ÷ 16 =	25 ÷ 17 =

0 ÷ 18 =	0 ÷ 19 =	0 ÷ 20 =	0 ÷ 21 =
1 ÷ 18 =	1 ÷ 19 =	1 ÷ 20 =	1 ÷ 21 =

2 ÷ 18 =	2 ÷ 19 =	2 ÷ 20 =	2 ÷ 21 =
3 ÷ 18 =	3 ÷ 19 =	3 ÷ 20 =	3 ÷ 21 =
4 ÷ 18 =	4 ÷ 19 =	4 ÷ 20 =	4 ÷ 21 =
5 ÷ 18 =	5 ÷ 19 =	5 ÷ 20 =	5 ÷ 21 =
6 ÷ 18 =	6 ÷ 19 =	6 ÷ 20 =	6 ÷ 21 =
7 ÷ 18 =	7 ÷ 19 =	7 ÷ 20 =	7 ÷ 21 =
8 ÷ 18 =	8 ÷ 19 =	8 ÷ 20 =	8 ÷ 21 =
9 ÷ 18 =	9 ÷ 19 =	9 ÷ 20 =	9 ÷ 21 =
10 ÷ 18 =	10 ÷ 19 =	10 ÷ 20 =	10 ÷ 21 =
11 ÷ 18 =	11 ÷ 19 =	11 ÷ 20 =	11 ÷ 21 =
12 ÷ 18 =	12 ÷ 19 =	12 ÷ 20 =	12 ÷ 21 =
13 ÷ 18 =	13 ÷ 19 =	13 ÷ 20 =	13 ÷ 21 =
14 ÷ 18 =	14 ÷ 19 =	14 ÷ 20 =	14 ÷ 21 =
15 ÷ 18 =	15 ÷ 19 =	15 ÷ 20 =	15 ÷ 21 =
16 ÷ 18 =	16 ÷ 19 =	16 ÷ 20 =	16 ÷ 21 =
17 ÷ 18 =	17 ÷ 19 =	17 ÷ 20 =	17 ÷ 21 =
18 ÷ 18 =	18 ÷ 19 =	18 ÷ 20 =	18 ÷ 21 =
19 ÷ 18 =	19 ÷ 19 =	19 ÷ 20 =	19 ÷ 21 =
20 ÷ 18 =	20 ÷ 19 =	20 ÷ 20 =	20 ÷ 21 =
21 ÷ 18 =	21 ÷ 19 =	21 ÷ 20 =	21 ÷ 21 =
22 ÷ 18 =	22 ÷ 19 =	22 ÷ 20 =	22 ÷ 21 =
23 ÷ 18 =	23 ÷ 19 =	23 ÷ 20 =	23 ÷ 21 =
24 ÷ 18 =	24 ÷ 19 =	24 ÷ 20 =	24 ÷ 21 =
25 ÷ 18 =	25 ÷ 19 =	25 ÷ 20 =	25 ÷ 21 =

0 ÷ 22 =	0 ÷ 23 =	0 ÷ 24 =	0 ÷ 25 =
1 ÷ 22 =	1 ÷ 23 =	1 ÷ 24 =	1 ÷ 25 =
2 ÷ 22 =	2 ÷ 23 =	2 ÷ 24 =	2 ÷ 25 =
3 ÷ 22 =	3 ÷ 23 =	3 ÷ 24 =	3 ÷ 25 =
4 ÷ 22 =	4 ÷ 23 =	4 ÷ 24 =	4 ÷ 25 =

5 ÷ 22 =	5 ÷ 23 =	5 ÷ 24 =	5 ÷ 25 =
6 ÷ 22 =	6 ÷ 23 =	6 ÷ 24 =	6 ÷ 25 =
7 ÷ 22 =	7 ÷ 23 =	7 ÷ 24 =	7 ÷ 25 =
8 ÷ 22 =	8 ÷ 23 =	8 ÷ 24 =	8 ÷ 25 =
9 ÷ 22 =	9 ÷ 23 =	9 ÷ 24 =	9 ÷ 25 =
10 ÷ 22 =	10 ÷ 23 =	10 ÷ 24 =	10 ÷ 25 =
11 ÷ 22 =	11 ÷ 23 =	11 ÷ 24 =	11 ÷ 25 =
12 ÷ 22 =	12 ÷ 23 =	12 ÷ 24 =	12 ÷ 25 =
13 ÷ 22 =	13 ÷ 23 =	13 ÷ 24 =	13 ÷ 25 =
14 ÷ 22 =	14 ÷ 23 =	14 ÷ 24 =	14 ÷ 25 =
15 ÷ 22 =	15 ÷ 23 =	15 ÷ 24 =	15 ÷ 25 =
16 ÷ 22 =	16 ÷ 23 =	16 ÷ 24 =	16 ÷ 25 =
17 ÷ 22 =	17 ÷ 23 =	17 ÷ 24 =	17 ÷ 25 =
18 ÷ 22 =	18 ÷ 23 =	18 ÷ 24 =	18 ÷ 25 =
19 ÷ 22 =	19 ÷ 23 =	19 ÷ 24 =	19 ÷ 25 =
20 ÷ 22 =	20 ÷ 23 =	20 ÷ 24 =	20 ÷ 25 =
21 ÷ 22 =	21 ÷ 23 =	21 ÷ 24 =	21 ÷ 25 =
22 ÷ 22 =	22 ÷ 23 =	22 ÷ 24 =	22 ÷ 25 =
23 ÷ 22 =	23 ÷ 23 =	23 ÷ 24 =	23 ÷ 25 =
24 ÷ 22 =	24 ÷ 23 =	24 ÷ 24 =	24 ÷ 25 =
25 ÷ 22 =	25 ÷ 23 =	25 ÷ 24 =	25 ÷ 25 =

ADDING NUMBERS

0 + 1 = 1	0 + 2 = 2	0 + 3 = 3	0 + 4 = 4	0 + 5 = 5
1 + 1 = 2	2 + 1 = 3	3 + 1 = 4	4 + 1 = 5	5 + 1 = 6
1 + 2 = 3	2 + 2 = 4	3 + 2 = 5	4 + 2 = 6	5 + 2 = 7
1 + 3 = 4	2 + 3 = 5	3 + 3 = 6	4 + 3 = 7	5 + 3 = 8
1 + 4 = 5	2 + 4 = 6	3 + 4 = 7	4 + 4 = 8	5 + 4 = 9
1 + 5 = 6	2 + 5 = 7	3 + 5 = 8	4 + 5 = 9	5 + 5 = 10
1 + 6 = 7	2 + 6 = 8	3 + 6 = 9	4 + 6 = 10	5 + 6 = 11
1 + 7 = 8	2 + 7 = 9	3 + 7 = 10	4 + 7 = 11	5 + 7 = 12
1 + 8 = 9	2 + 8 = 10	3 + 8 = 11	4 + 8 = 12	5 + 8 = 13
1 + 9 = 10	2 + 9 = 11	3 + 9 = 12	4 + 9 = 13	5 + 9 = 14
1 + 10 = 11	2 + 10 = 12	3 +10 = 13	4 + 10 = 14	5 + 10 = 15
1 + 11 = 12	2 + 11 = 13	3 +11 = 14	4 + 11 = 15	5 + 11= 16
1 + 12 = 13	2 + 12 = 14	3 +12 = 15	4 + 12 = 16	5 + 12 = 17
1 + 13 = 14	2 + 13 = 15	3 +13 = 16	4 + 13 = 17	5 + 13 = 18
1 + 14 = 15	2 + 14 = 16	3 + 14 = 17	4 + 14 = 18	5 + 14 = 19
1 + 15 = 16	2 + 15 = 17	3 + 15 = 18	4 + 15 = 19	5 + 15 = 20
1 + 16 = 17	2 + 16 = 18	3 + 16 = 19	4 + 16 = 20	5 + 16 = 21
1 + 17 = 18	2 + 17 = 19	3 + 17 = 20	4 + 17 = 21	5 + 17 = 22
1 + 18 = 19	2 + 18 = 20	3 + 18 = 21	4 + 18 = 22	5 + 18 = 23
1 + 19 = 20	2 + 19 = 21	3 + 19 = 22	4 + 19 = 23	5 + 19 = 24

1 + 20 = 21	2 +20 = 22	3 + 20 = 23	4 + 20 = 24	5 + 20 = 25
1 + 21= 22	2+ 21 = 23	3 + 21 = 24	4 + 21 = 25	5 + 21 = 26
1 + 22 = 23	2 +22 = 24	3 + 22 = 25	4 + 22 = 26	5 + 22 = 27
1 + 23 = 24	2 +23 = 25	3 + 23 = 26	4 + 23 = 27	5 + 23 = 28
1 + 24 = 25	2 + 24 = 26	3 + 24 = 27	4 + 24 = 28	5 + 24 = 29
1 + 25 = 26	2 + 25 = 27	3 + 25 = 28	4 + 25 = 29	5 + 25 = 30

6 + 0 = 6	7 + 0 = 7	8 + 0 = 8	9 + 0 = 9	10 + 0 = 10
6 + 1 = 7	7 + 1 = 8	8 + 1 = 9	9 +1 = 10	10 + 1 = 11
6 + 2 = 8	7 + 2 = 9	8 + 2 = 10	9 + 2 = 11	10 + 2 = 12
6 + 3 = 9	7 + 3 = 10	8 + 3 = 11	9 + 3 = 12	10 + 3 = 13
6 + 4 = 10	7 + 4 = 11	8 + 4 = 12	9 + 4 = 13	10 + 4 = 14
6 + 5 = 11	7 + 5 = 12	8 + 5 = 13	9 + 5 = 14	10 + 5 = 15
6 + 6 = 12	7 + 6 = 13	8 + 6 = 14	9 + 6 = 15	10 + 6 = 16
6 + 7 = 13	7 + 7 = 14	8 + 7 = 15	9 + 7 = 16	10 + 7 = 17
6 + 8 = 14	7 + 8 = 15	8 + 8 = 16	9 + 8 = 17	10 + 8 = 18
6 + 9 = 15	7 + 9 = 16	8 + 9 = 17	9 + 9 = 18	10 + 9 = 19
6 + 10 = 16	7 + 10 = 17	8 + 10 = 18	9 + 10 = 19	10 + 10 = 20
6 + 11 = 17	7 + 11 = 18	8 + 11 = 19	9 + 11 = 20	10 + 11 = 21
6 + 12 = 18	7 + 12 = 19	8 + 12 = 20	9 + 12 = 21	10 + 12 = 22
6 + 13 = 19	7 + 13 = 20	8 + 13 = 21	9 + 13 = 22	10 + 13 = 23
6 + 14 = 20	7 + 14 = 21	8 + 14 = 22	9 + 14 = 23	10 + 14 = 24
6 + 15 = 21	7 + 15 = 22	8 + 15 = 23	9 + 15 = 24	10 + 15 = 25
6 + 16 = 22	7 + 16 = 23	8 + 16 = 24	9 + 16 = 25	10 + 16 = 26
6 + 17 = 23	7 + 17 = 24	8 + 17 = 25	9 + 17 = 26	10 + 17 = 27
6 + 18 = 24	7 + 18 = 25	8 + 18 = 26	9 + 18 = 27	10 + 18 = 28
6 + 19 = 25	7 + 19 = 26	8 + 19 = 27	9 + 19 = 28	10 + 19 = 29
6 + 20 = 26	7 + 20 = 27	8 + 20 = 28	9 + 20 = 29	10 + 20 = 30
6 + 21 = 27	7 + 21 = 28	8 + 21 = 29	9 + 21 = 30	10 + 21 = 31
6 + 22 = 28	7 + 22 = 29	8 + 22 = 30	9 + 22 = 31	10 + 22 = 32

6 + 23 = 29	7 + 23 = 30	8 + 23 = 31	9 + 23 = 32	10 + 23 = 33
6 + 24 = 30	7 + 24 = 31	8 + 24 = 32	9 + 24 = 33	10 + 24 = 34
6 + 25 = 31	7 + 25 = 32	8 + 25 = 33	9 + 25 = 34	10 + 25 = 35

11 + 0 = 11	12 + 0 = 12	13 + 0 = 13	14 + 0 = 14	15 + 0 = 15
11 + 1 = 12	12 + 1 = 13	13 + 1 = 14	14 + 1 = 15	15 + 1 = 16
11 + 2 = 13	12 + 2 = 14	13 + 2 = 15	14 + 2 = 16	15 + 2 = 17
11 + 3 = 14	12 + 3 = 15	13 + 3 = 16	14 + 3 = 17	15 + 3 = 18
11 + 4 = 15	12 + 4 = 16	13 + 4 = 17	14 + 4 = 18	15 + 4 = 19
11 + 5 = 16	12 + 5 = 17	13 + 5 = 18	14 + 5 = 19	15 + 5 = 20
11 + 6 = 17	12 + 6 = 18	13 + 6 = 19	14 + 6 = 20	15 + 6 = 21
11 + 7 = 18	12 + 7 = 19	13 + 7 = 20	14 + 7 = 21	15 + 7 = 22
11 + 8 = 19	12 + 8 = 20	13 + 8 = 21	14 + 8 = 22	15 + 8 = 23
11 + 9 = 20	12 + 9 = 21	13 + 9 = 22	14 + 9 = 23	15 + 9 = 24
11 + 10 = 21	12 + 10 = 22	13 + 10 = 23	14 + 10 = 24	15 + 10 = 25
11 + 11 = 22	12 + 11 = 23	13 + 11 = 24	14 + 11 = 25	15 + 11 = 26
11 + 12 = 23	12 + 12 = 24	13 + 12 = 25	14 + 12 = 26	15 + 12 = 27
11+ 13 = 24	12 + 13 = 25	13 + 13 = 26	14 + 13 = 27	15 + 13 = 28
11+ 14 = 25	12 + 14 = 26	13 + 14 = 27	14 + 14 = 28	15 + 14 = 29
11+ 15 = 26	12 + 15 = 27	13 + 15 = 28	14 + 15 = 29	15 + 15 = 30
11 + 16 = 27	12 + 16 = 28	13 + 16 = 29	14 + 16 = 30	15 + 16 = 31
11 + 17 = 28	12 + 17 = 29	13 + 17 = 30	14 + 17 = 31	15 + 17 = 32
11 + 18 = 29	12 + 18 = 30	13 + 18 = 31	14 + 18 = 32	15 + 18 = 33
11 + 19 = 30	12 + 19 = 31	13 + 19 = 32	14 +19 = 33	15 + 19 = 34
11 + 20 = 31	12 + 20 = 32	13 + 20 = 33	14 + 20 = 34	15 + 20 = 35
11 + 21 = 32	12 + 21 = 33	13 +21 = 34	14 + 21 = 35	15 + 21 = 36
11 + 22 = 33	12 + 22 = 34	13 + 22 = 35	14 + 22 = 36	15 + 22 = 37
11 + 23 = 34	12 + 23 = 35	13 + 23 = 36	14 + 23 = 37	15 + 23 = 38
11 + 24 = 35	12 + 24 = 36	13 + 24 = 37	14 + 24 = 38	15 + 24 = 39
11 + 25 = 36	12 + 25 = 37	13 + 25 = 38	14 + 25 = 39	15 + 25 = 40

16 + 0 = 16	17 + 0 = 17	18 + 0 = 18	19 + 0 = 19	20 + 0 = 20
16 + 1 = 17	17 + 1 = 18	18 + 1 = 19	19 + 1 = 20	20 + 1 = 21
16 + 2 = 18	17 + 2 = 19	18 + 2 = 20	19 + 2 = 21	20 + 2 = 22
16 + 3 = 19	17 + 3 = 20	18 + 3 = 21	19 + 3 = 22	20 + 3 = 23
16 + 4 = 20	17 + 4 = 21	18 + 4 = 22	19 + 4 = 23	20 + 4 = 24
16 + 5 = 21	17 + 5 = 22	18 + 5 = 23	19 + 5 = 24	20 + 5 = 25
16 + 6 = 22	17 + 6 = 23	18 + 6 = 24	19 +6 = 25	20 + 6 = 26
16 + 7 = 23	17 + 7 = 24	18 +7 = 25	19 + 7 = 26	20 + 7 = 27
16 + 8 = 24	17 + 8 = 25	18 + 8 = 26	19 + 8 = 27	20 + 8 = 28
16 + 9 = 25	17 + 9 = 26	18 + 9 = 27	19 + 9 = 28	20 + 9 = 29
16 + 10 = 26	17 + 10 = 27	18 + 10 = 28	19 + 10 = 29	20 + 10 = 30
16 + 11 = 27	17 + 11 = 28	18 + 11 = 29	19 + 11 = 30	20 + 11 = 31
16 + 12 = 28	17 + 12 = 29	18 + 12 = 30	19 + 12 = 31	20 + 12 = 32
16 + 13 = 29	17 + 13 = 30	18 +13 = 31	19 +13 = 32	20 + 13 = 33
16 + 14 = 30	17 + 14 = 31	18 +14 = 32	19 + 14 = 33	20 + 14 = 34
16 + 15 = 31	17 + 15 = 32	18 +15 = 33	19 + 15 = 34	20 + 15 = 35
16 + 16 = 32	17 + 16 = 33	18 + 16 = 34	19 +16 = 35	20 + 16 = 36
16 + 17 = 33	17 + 17 = 34	18 + 17 = 35	19 + 17 = 36	20 + 17 = 37
16 +18 = 34	17 + 18 = 35	18 + 18 = 36	19 + 18 = 37	20 + 18 = 38
16 +19 = 35	17 + 19 = 36	18 + 19 = 37	19 + 19 = 38	20 + 19 = 39
16 + 20 = 36	17 + 20 = 37	18 + 20 = 38	19 + 20 = 39	20 + 20 = 40
16 + 21 = 37	17 + 21 = 38	18 + 21 = 39	19 + 21 = 40	20 + 21 = 41
16 + 22 = 38	17 + 22 = 39	18 + 22 = 40	19 + 22 = 41	20 + 22 = 42
16 + 23 = 39	17 + 23 = 40	18 + 23 = 41	19 + 23 = 42	20 + 23 = 43
16 + 24 = 40	17 + 24 = 41	18 + 24 = 42	19 + 24 = 43	20 + 24 = 44
16 + 25 = 41	17 + 25 = 42	18 + 25 = 43	19 + 25 = 44	20 + 25 = 45

21 + 0 = 21	22 + 0 = 22	23 + 0 = 23	24 + 0 = 24	25 + 0 = 25
21 + 1 = 22	22 + 1 = 23	23 + 1 = 24	24 + 1 = 25	25 + 1 = 26

21 + 2 = 23	22 + 2 = 24	23 + 2 = 25	24 + 2 = 26	25 + 2 = 27
21 + 3 = 24	22 + 3 = 25	23 + 3 = 26	24 + 3 = 27	25 + 3 = 28
21 + 4 = 25	22 + 4 = 26	23 + 4 = 27	24 + 4 = 28	25 + 4 = 29
21 + 5 = 26	22 +5 = 27	23 + 5 = 28	24 + 5 = 29	25 + 5 = 30
21 + 6 = 27	22 + 6 = 28	23 + 6 = 29	24 + 6 = 30	25 + 6 = 31
21 + 7 = 28	22 + 7 = 29	23 + 7 = 30	24 + 7 = 31	25 + 7 = 32
21 + 8 = 29	22 + 8 = 30	23 + 8 = 31	24 + 8 = 32	25 + 8 = 33
21 + 9 = 30	22 + 9 = 31	23 + 9 = 32	24 + 9 = 33	25 + 9 = 34
21 + 10 = 31	22 + 10 = 32	23 +10 = 33	24 +10 = 34	25 + 10 = 35
21 + 11 = 32	22 + 11= 33	23 + 11= 34	24 + 11= 35	25 + 11= 36
21 + 12 = 33	22 + 12 = 34	23 +12 = 35	24 + 12 = 36	25 + 12 = 37
21 + 13 = 34	22 + 13 = 35	23 + 13 = 36	24 + 13 = 37	25 + 13 = 38
21 + 14 = 35	22 + 14 = 36	23 + 14 = 37	24 + 14 = 38	25 + 14 = 39
21 + 15 = 36	22 + 15 = 37	23 + 15 = 38	24 + 15 = 39	25 + 15 = 40
21 + 16 = 37	22 + 16 = 38	23 + 16 = 39	24 + 16 = 40	25 + 16 = 41
21 + 17 = 38	22 + 17 = 39	23 + 17= 40	24 + 17 = 41	25 + 17 = 42
21 + 18 = 39	22 + 18 = 40	23 + 18 = 41	24 + 18 = 42	25 + 18 = 43
21 + 19 = 40	22 + 19 = 41	23 + 19 = 42	24 + 19 = 43	25 + 19 = 44
21 + 20 = 41	22 + 20 = 42	23 + 20 = 43	24 + 20 = 44	25 + 20 = 45
21 + 21 = 42	22 + 21 = 43	23 + 21 = 44	24 + 21 = 45	25 +21 = 46
21 + 22 = 43	22 + 22 = 44	23 + 22 = 45	24 + 22 = 46	25 + 22 = 47
21 + 23 = 44	22 + 23 = 45	23 + 23 = 46	24 + 23 = 47	25 + 23 = 48
21 + 24 = 45	22 + 24 = 46	23 + 24 = 47	24 + 24 = 48	25 + 24 = 49
21 + 25 = 46	22 + 25 = 47	23 + 25 = 48	24 + 25 = 49	25 + 25 = 50

SUBTRACTING NUMBERS

1- 0 = 1	2 - 0 = 2	3 - 0 = 3	4 - 0 = 4	5 – 0 = 5
1 - 1 = 0	2 - 1 = 1	3 - 1 = 2	4 - 3 = 1	5 - 1 = 4
2 - 1 = 1	2 - 2 = 0	3 – 2 = 1	4 – 2 = 2	5 - 2 = 3
3 - 1 = 2	3 - 2 = 1	3 – 3 = 0	4 – 3 = 1	5 - 3 = 2
4 - 1 = 3	4 - 2 = 2	4 - 3 = 1	4 – 4 = 0	5 – 4 = 1
5 – 1 = 4	5 – 2 = 3	5 – 3 = 2	5 - 4 = 1	5 - 5 = 0
6 – 1 = 5	6 - 2 = 4	6 – 3 = 3	6 – 4 = 2	6 – 5 = 1
7 – 1 = 6	7 – 2 = 5	7 – 3 = 4	7 – 4 = 3	7 – 5 = 2
8 – 1 = 7	8 - 2 = 6	8 - 3 = 5	8 – 4 = 4	8 - 5 = 3
9 - 1 = 8	9 - 2 = 7	9 - 3 = 6	9 - 4 = 5	9 - 5 = 4
10 – 1 = 9	10 - 2 = 8	10 - 3 = 7	10 – 4 = 6	10 - 5 = 5
11 – 1 = 10	11 - 2 = 9	11 - 3 = 8	11 - 4 = 7	11 - 5 = 6
12 – 1 = 11	12 - 2 = 10	12 - 3 = 9	12 - 4 = 8	12 - 5 = 7
13 – 1 = 12	13 – 2 = 11	13 - 3 = 10	13 - 4 = 9	13 - 5 = 8
14 – 1 = 13	14 - 2 = 12	14 - 3 = 11	14 - 4 = 10	14 - 5 = 9
15 - 1 = 14	15 - 2 = 13	15 - 3 = 12	15 - 4 = 11	15 - 5 = 10
16 - 1 = 15	16 - 2 = 14	16 - 3 = 13	16 - 4 = 12	16 - 5 = 11
17 – 1 = 16	17 – 2 = 15	17 – 3 = 14	17 – 4 = 13	17 – 5 = 12
18 – 1 = 17	18 – 2 = 16	18 – 3 = 15	18 – 4 = 14	18 – 5 = 13
19 - 1 = 18	19 - 2 = 17	19 - 3 = 16	19 - 4 = 15	19 - 5 = 14

20 - 1 = 19	20 - 2 = 18	20 - 3 = 17	20 - 4 = 16	20 - 5 = 15
21 - 1 = 20	21- 2 = 19	21 - 3 = 18	21- 4 = 17	21- 5 = 16
22 - 1 = 21	22 -2 = 20	22 - 3 = 19	22 - 4 = 18	22 - 5 = 17
23 - 1 = 22	23 - 2 = 21	23 - 3 = 20	23 - 4 = 19	23 - 5 = 18
24 - 1 = 23	24 - 2 = 22	24 - 3 = 21	24 - 4 = 20	24 - 5 = 19
25 - 1 = 24	25 - 2 = 23	25 - 3 = 22	25 - 4 = 21	25 - 5 = 20

6 - 0 = 6	7 - 0 = 7	8 - 0 = 8	9 - 0 = 9	10 - 0 = 10
6 - 1 = 5	7 - 1 = 6	8 - 1 = 7	9 - 1 = 8	10 - 1 = 9
6 - 2 = 4	7 - 2 = 5	8 - 2 = 6	9 - 2 = 7	10 - 2 = 8
6 - 3 = 3	7 - 3 = 4	8 - 3 = 5	9 - 3 = 6	10 - 3 = 7
6- 4 = 2	7 - 4 = 3	8 - 4 = 4	9 - 4 = 5	10 - 4 = 6
6 - 5 = 1	7- 5 = 2	8 - 5 = 3	9 - 5 = 4	10 - 5 = 5
6 - 6 = 0	7 - 6 = 1	8 - 6 = 2	9 - 6 = 3	10 - 6 = 4
7 – 6 = 1	7 – 7 = 0	8 – 7 = 1	9 – 7 = 2	10 – 7 = 3
8 – 6 = 2	8 - 7 = 1	8 – 8 = 0	9 - 8 = 1	10 – 8 = 2
9 - 6 = 3	9 – 7 = 2	9 - 8 = 1	9 - 9 = 0	10 – 9 = 1
10 - 6 = 4	10 - 7 = 3	10 - 8 = 2	10 - 9 = 1	10 - 10 = 0
11- 6 = 5	11- 7 = 4	11 - 8 = 3	11- 9 = 2	11 - 10 = 1
12 - 6 = 6	12 - 7 = 5	12 - 8 = 4	12 - 9 = 3	12 - 10 = 2
13 - 6 = 7	13 - 7 = 6	13 - 8 = 5	13 - 9 = 4	13 -10 = 3
14 - 6 = 8	14 - 7 = 7	14 - 8 = 6	14 - 9 = 5	14 - 10 = 4
15 - 6 = 9	15 - 7 = 8	15- 8 = 7	15 - 9 = 6	15 - 10 = 5
16 - 6 = 10	16 - 7 = 9	16 - 8 = 8	16 - 9 = 7	16 - 10 = 6
17 - 6 = 11	17 - 7 = 10	17 – 8 = 9	17- 9 = 8	17 - 10 = 7
18 - 6 = 12	18 - 7 = 11	18 - 8 = 10	18 - 9 = 9	18 - 10 = 8
19 - 6 = 13	19 - 7 = 12	19 - 8 = 11	19 - 9 = 10	19 - 10 = 9
20 - 6 = 14	20 - 7 = 13	20 - 8 = 12	20 - 9 = 11	20 - 10 = 10
21 - 6 = 15	21 - 7 = 14	21- 8 = 13	21 - 9 = 12	21 - 10 = 11
22 - 6 = 16	22 - 7 = 15	22 - 8 = 14	22 - 9 = 13	22 - 10 = 12

23 - 6 = 17	23 - 7 = 16	23 - 8 = 15	23 - 9 = 14	23 - 10 = 13
24 - 6 = 18	24 - 7 = 17	24 - 8 = 16	24 - 9 = 15	24 - 10 = 14
25 - 6 = 19	25 - 7 = 18	25 - 8 = 17	25 - 9 = 16	25 - 10 = 15

11 - 0 = 11	12 - 0 = 12	13 - 0 = 13	14 - 0 = 14	15 - 0 = 15
11- 1 = 10	12 - 1 = 11	13 - 1 = 12	14 - 1 = 13	15 - 1 = 14
11- 2 = 9	12 - 2 = 10	13 - 2 = 11	14 - 2 = 12	15 - 2 = 13
11- 3 = 8	12 - 3 = 9	13 - 3 = 10	14 - 3 = 11	15 - 3 = 12
11- 4 = 7	12 - 4 = 8	13 - 4 = 9	14 - 4 = 10	15 - 4 = 11
11- 5 = 6	12 - 5 = 7	13 - 5 = 8	14 - 5 = 9	15 - 5 = 10
11- 6 = 5	12 - 6 = 6	13 - 6 = 7	14 - 6 = 8	15 - 6 = 9
11- 7 = 4	12 - 7 = 5	13 - 7 = 6	14 - 7 = 7	15 - 7 = 8
11- 8 = 3	12 - 8 = 4	13 - 8 = 5	14 - 8 = 6	15 - 8 = 7
11- 9 = 2	12 - 9 = 3	13 - 9 = 4	14 - 9 = 5	15 - 9 = 6
11- 10 = 1	12 - 10 = 2	13 - 10 = 3	14 - 10 = 4	15 - 10 = 5
11- 11 = 0	12 - 11 = 1	13 - 11 = 2	14 - 11 = 3	15 - 11 = 4
12 - 11 = 1	12 - 12 = 0	13 - 12 = 1	14 - 12 = 2	15 - 12 = 3
13 - 11 = 2	13 - 12 = 1	13 - 13 = 0	14 - 13 = 1	15 - 13 = 2
14 - 11 = 3	14 - 12 = 2	14 - 13 = 1	14 - 14 = 0	15 - 14 = 1
15 – 11 = 4	15 – 12 = 3	15 – 13 = 2	15 – 14 = 1	15 – 15 = 0
16 - 11 = 5	16 - 12 = 4	16 - 13 = 3	16 - 14 = 2	16 - 15 = 1
17 - 11 = 6	17- 12 = 5	17 - 13 = 4	17 - 14 = 3	17 - 15 = 2
18 - 11 = 7	18 - 12 = 6	18 - 13 = 5	18 - 14 = 4	18 - 15 = 3
19 - 11 = 8	19 - 12 = 7	19 - 13 = 6	19 - 14 = 5	19 - 15 = 4
20 - 11 = 9	20 - 12 = 8	20 - 13 = 7	20 - 14 = 6	20 - 15 = 5
21 – 11 = 10	21 – 12 = 9	21 – 13 = 8	21 – 14 = 7	21 – 15 = 6
22 - 11 = 11	22 - 12 = 10	22 - 13 = 9	22 - 14 = 8	22 - 15 = 7
23 - 11 = 12	23 - 12 = 11	23 - 13 = 10	23 - 14 = 9	23 -15 = 8
24 - 11 = 13	24 - 12 = 12	24 - 13 = 11	24 - 14 = 10	24 - 15 = 9
25 - 11 = 14	25 - 12 = 13	25 - 13 = 12	25 - 14 = 11	25 - 15 = 10

16 - 0 = 16	17 - 0 = 17	18 - 0 = 18	19 - 0 = 19	20 - 0 = 20
16 - 1 = 15	17- 1 = 16	18 - 1 = 17	19 - 1 = 18	20 - 1 = 19
16 - 2 = 14	17- 2 = 15	18 - 2 = 16	19 - 2 = 17	20 - 2 = 18
16 - 3 = 13	17- 3 = 14	18 - 3 = 15	19 - 3 = 16	20 - 3 = 17
16 - 4 = 12	17 - 4 = 13	18 - 4 = 14	19 - 4 = 15	20 - 4 = 16
16 - 5 = 11	17 - 5 = 12	18 - 5 = 13	19 - 5 = 14	20 - 5 = 15
16 - 6 = 10	17 - 6 = 11	18 - 6 = 12	19 - 6 = 13	20 - 6 = 14
16 - 7 = 9	17 - 7 = 10	18 - 7 = 11	19 - 7 = 12	20 - 7 = 13
16 - 8 = 8	17 - 8 = 9	18 - 8 = 10	19 - 8 = 11	20 - 8 = 12
16 - 9 = 7	17 - 9 = 8	18 - 9 = 9	19 - 9 = 10	20 - 9 = 11
16 -10 = 6	17 - 10 = 7	18 - 10 = 8	19 - 10 = 9	20 - 10 = 10
16 - 11 = 5	17 - 11 = 6	18 - 11 = 7	19 - 11 = 8	20 - 11 = 9
16 -12 = 4	17 - 12 = 5	18 - 12 = 6	19 - 12 = 7	20 - 12 = 8
16 - 13 = 3	17 - 13 = 4	18 - 13 = 5	19 - 13 = 6	20 - 13 = 7
16 - 14 = 2	17 -14 = 3	18 - 14 = 4	19 - 14 = 5	20 - 14 = 6
16 - 15 = 1	17- 15 = 2	18 - 15 = 3	19 - 15 = 4	20 - 15 = 5
16 - 16 = 0	17 - 16 = 1	18 - 16 = 2	19 - 16 = 3	20 - 16 = 4
17 - 16 = 1	17 - 17 = 0	18 - 17 = 1	19 - 17 = 2	20 - 17 = 3
18 - 16 = 2	18 - 17 = 1	18 - 18 = 0	19 - 18 = 1	20 - 18 = 2
19 - 16 = 3	19 - 17 = 2	18 - 18 = 1	19 - 19 = 0	20 - 19 = 1
20 - 16 = 4	20 - 17 = 3	20 - 18 = 2	20 - 19 = 1	20 - 20 = 0
21 - 16 = 5	21 - 17 = 4	21- 18 = 3	21- 19 = 2	21- 20 = 1
22 - 16 = 6	22 - 17 = 5	22 - 18 = 4	22- 19 = 3	22 - 20 = 2
23 - 16 = 7	23 - 17 = 6	23 - 18 = 5	23 - 19 = 4	23 - 20 = 3
24 - 16 = 8	24 - 17 = 7	24 - 18 = 6	24 - 19 = 5	24 - 20 = 4
25 - 16 = 9	25 -17 = 8	25 - 18 = 7	25 - 19 = 6	25 - 20 = 5

21 - 0 = 21	22 - 0 = 22	23 - 0 = 23	24 - 0 = 24	25 - 0 = 25
21 - 1 = 20	22 - 1 = 21	23 - 1 = 22	24 - 1 = 23	25 - 1 = 24
21 - 2 = 19	22 - 2 = 20	23 - 2 = 21	24 - 2 = 22	25 - 2 = 23

21 - 3 = 18	22 - 3 = 19	23 - 3 = 20	24 - 3 = 21	25 – 3 = 22
21 - 4 = 17	22 - 4 = 18	23 - 4 = 19	24 - 4 = 20	25 – 4 = 21
21 - 5 = 16	22 - 5 = 17	23 - 5 = 18	24 - 5 = 19	25 – 5 = 20
21 - 6 = 15	22 - 6 = 16	23 - 6 = 17	24 - 6 = 18	25 – 6 = 19
21 - 7 = 14	22 - 7 = 15	23 - 7 = 16	24 - 7 = 17	25 – 7 = 18
21 - 8 = 13	22 - 8 = 14	23 - 8 = 15	24 - 8 = 16	25 – 8 = 17
21 - 9 = 12	22 - 9 = 13	23 - 9 = 14	24 - 9 = 15	25 – 9 = 16
21 - 10 = 11	22 - 10 = 12	23 - 10 = 13	24 - 10 = 14	25 – 10 = 15
21 - 11 = 10	22 - 11 = 11	23 - 11 = 12	24 - 11 = 13	25 – 11 = 14
21 - 12 = 9	22 - 12 = 10	23 - 12 = 11	24 - 12 = 12	25 – 12 = 13
21- 13 = 8	22 - 13 = 9	23 - 13 = 10	24 - 13 = 11	25 – 13 = 12
21 - 14 = 7	22 - 14 = 8	23 - 14 = 9	24 - 14 = 10	25 – 14 = 11
21 - 15 = 6	22 - 15 = 7	23 -15 = 8	24 - 15 = 9	25 – 15 = 10
21 - 16 = 5	22 - 16 = 6	23 - 16 = 7	24 - 16 = 8	25 – 16 = 9
21 - 17 = 4	22 - 17 = 5	23 - 17 = 6	24 - 17 = 7	25 – 17 = 8
21 - 18 = 3	22 - 18 = 4	23 - 18 = 5	24 - 18 = 6	25 – 18 = 7
21 - 19 = 2	22 - 19 = 3	23 - 19 = 4	24 - 19 = 5	25 – 19 = 6
21 - 20 = 1	22 - 20 = 2	23 - 20 = 3	24 - 20 = 4	25 – 20 = 5
21 - 21 = 0	22 - 21 = 1	23 - 21 = 2	24 - 21 = 3	25 – 21 = 4
22 - 21 = 1	22 - 22 = 0	23 - 22 = 1	24 - 22 = 2	25 – 22 = 3
23 - 21 = 2	23- 22 = 1	23 - 23 = 0	24 - 23 = 1	25 – 23 = 2
24 - 21 = 3	24 - 22 = 2	24 - 23 = 1	24 - 24 = 0	25 – 24 = 1
25 - 21 = 4	25 - 22 = 3	25 - 23 = 2	24 - 23 = 1	25 – 25 = 0

MULTIPLYING NUMBERS

1 x 0 = 0	2 x 0 = 0	3 x 0 = 0	4 x 0 = 0	5 x 0 = 0
1 x 1 = 1	2 x 1 = 2	3 x 1 = 3	4 x 1 = 4	5 x 1 = 5
1 x 2 = 2	2 x 2 = 4	3 x 2 = 6	4 x 2 = 8	5 x 2 = 10
1 x 3 = 3	2 x 3 = 6	3 x 3 = 9	4 x 3 = 12	5 x 3 = 15
1 x 4 = 4	2 x 4 = 8	3 x 4 = 12	4 x 4 = 16	5 x 4 = 20
1 x 5 = 5	2 x 5 = 10	3 x 5 = 15	4 x 5 = 20	5 x 5 = 25
1 x 6 = 6	2 x 6 = 12	3 x 6 = 18	4 x 6 = 24	5 x 6 = 30
1 x 7 = 7	2 x 7 = 14	3 x 7 = 21	4 x 7 = 28	5 x 7 = 35
1 x 8 = 8	2 x 8 = 16	3 x 8 = 24	4 x 8 = 32	5 x 8 = 40
1 x 9 = 9	2 x 9 = 18	3 x 9 = 27	4 x 9 = 36	5 x 9 = 45
1 x 10 = 10	2 x 10 = 20	3 x 10 = 30	4 x 10 = 40	5 x 10 = 50
1 x 11 = 11	2 x 11 = 22	3 x 11 = 33	4 x 11 = 44	5 x 11 = 55
1 x 12 = 12	2 x 12 = 24	3 x 12 = 36	4 x 12 = 48	5 x 12 = 60
1 x 13 = 13	2 x 13 = 26	3 x 13 = 39	4 x 13 = 52	5 x 13 = 65
1 x 14 = 14	2 x 14 = 28	3 x 14 = 42	4 x 14 = 56	5 x 14 = 70
1 x 15 = 15	2 x 15 = 30	3 x 15 = 45	4 x 15 = 60	5 x 15 = 75
1 x 16 = 16	2 x 16 = 32	3 x 16 = 48	4 x 16 = 64	5 x 16 = 80
1 x 17 = 17	2 x 17 = 34	3 x 17 = 51	4 x 17 = 68	5 x 17 = 85
1 x 18 = 18	2 x 18 = 36	3 x 18 = 54	4 x 18 = 72	5 x 18 = 90
1 x 19 = 19	2 x 19 = 38	3 x 19 = 57	4 x 19 = 76	5 x 19 = 95

1 x 20 = 20	2 x 20 = 40	3 x 20 = 60	4 x 20 = 80	5 x 20 = 100
1 x 21 = 21	2 x 21 = 42	3 x 21 = 63	4 x 21 = 84	5 x 21 = 105
1 x 22 = 22	2 x 22 = 44	3 x 22 = 66	4 x 22 = 88	5 x 22 = 110
1 x 23 = 23	2 x 23 = 46	3 x 23 = 69	4 x 23 = 92	5 x 23 = 115
1 x 24 = 24	2 x 24 = 48	3 x 24 = 72	4 x 24 = 96	5 x 24 = 120
1 x 25 = 25	2 x 25 = 50	3 x 15 = 75	4 x 25 = 100	5 x 25 = 125

6 x 0 = 0	7 x 0 = 0	8 x 0 = 0	9 x 0 = 0
6 x 1 = 6	7 x 1 = 7	8 x 1 = 8	9 x 1 = 9
6 x 2 = 12	7 x 2 = 14	8 x 2 = 16	9 x 2 = 18
6 x 3 = 18	7 x 3 = 21	8 x 3 = 24	9 x 3 = 27
6 x 4 = 24	7 x 4 = 28	8 x 4 = 32	9 x 4 = 36
6 x 5 = 30	7 x 5 = 35	8 x 5 = 40	9 x 5 = 45
6 x 6 = 36	7 x 6 = 42	8 x 6 = 48	9 x 6 = 54
6 x 7 = 42	7 x 7 = 49	8 x 7 = 56	9 x 7 = 63
6 x 8 = 48	7 x 8 = 56	8 x 8 = 64	9 x 8 = 72
6 x 9 = 54	7 x 9 = 63	8 x 9 = 72	9 x 9 = 81
6 x 10 = 60	7 x 10 = 70	8 x 10 = 80	9 x 10 = 90
6 x 11 = 66	7 x 11 = 77	8 x 11 = 88	9 x 11 = 99
6 x 12 = 72	7 x 12 = 84	8 x 12 = 96	9 x 12 = 108
6 x 13 = 78	7 x 13 = 91	8 x 13 = 104	9 x 13 = 117
6 x 14 = 84	7 x 14 = 98	8 x 14 = 112	9 x 14 = 126
6 x 15 = 90	7 x 15 = 105	8 x 15 = 120	9 x 15 = 135
6 x 16 = 96	7 x 16 = 112	8 x 16 = 128	9 x 16 = 144
6 x 17 = 102	7 x 17 = 119	8 x 17 = 136	9 x 17 = 153
6 x 18 = 108	7 x 18 = 126	8 x 18 = 144	9 x 18 = 162
6 x 19 = 114	7 x 19 = 133	8 x 19 = 152	9 x 19 = 171
6 x 20 = 120	7 x 20 = 140	8 x 20 = 160	9 x 20 = 180
6 x 21 = 126	7 x 21 = 147	8 x 21 = 168	9 x 21 = 189
6 x 22 = 132	7 x 22 = 154	8 x 22 = 176	9 x 22 = 198

6 x 23 = 138	7 x 23 = 161	8 x 23 = 184	9 x 23 = 207
6 x 24 = 144	7 x 24 = 168	8 x 24 = 192	9 x 24 = 216
6 x 25 = 150	7 x 25 = 175	8 x 25 = 200	9 x 25 = 225

10 x 0 = 0	11 x 0 = 0	12 x 0 = 0	13 x 0 = 0
10 x 1 = 10	11 x 1 = 11	12 x 1 = 12	13 x 1 = 13
10 x 2 = 20	11 x 2 = 22	12 x 2 = 24	13 x 2 = 26
10 x 3 = 30	11 x 3 = 33	12 x 3 = 36	13 x 3 = 39
10 x 4 = 40	11 x 4 = 44	12 x 4 = 48	13 x 4 = 52
10 x 5 = 50	11 x 5 = 55	12 x 5 = 60	13 x 5 = 65
10 x 6 = 60	11 x 6 = 66	12 x 6 = 72	13 x 6 = 78
10 x 7 = 70	11 x 7 = 77	12 x 7 = 84	13 x 7 = 91
10 x 8 = 80	11 x 8 = 88	12 x 8 = 96	13 x 8 = 104
10 x 9 = 90	11 x 9 = 99	12 x 9 = 108	13 x 9 = 117
10 x 10 = 100	11 x 10 = 110	12 x 10 = 120	13 x 10 = 130
10 x 11 = 110	11 x 11 = 121	12 x 11 = 132	13 x 11 = 143
10 x 12 = 120	11 x 12 = 132	12 x 12 = 144	13 x 12 = 156
10 x 13 = 130	11 x 13 = 143	12 x 13 = 156	13 x 13 = 169
10 x 14 = 140	11 x 14 = 154	12 x 14 = 168	13 x 14 = 182
10 x 15 = 150	11 x 15 = 165	12 x 15 = 180	13 x 15 = 195
10 x 16 = 160	11 x 16 = 176	12 x 16 = 192	13 x 16 = 208
10 x 17 = 170	11 x 17 = 187	12 x 17 = 204	13 x 17 = 221
10 x 18 = 180	11 x 18 = 198	12 x 18 = 216	13 x 18 = 234
10 x 19 = 190	11 x 19 = 209	12 x 19 = 228	13 x 19 = 247
10 x 20 = 200	11 x 20 = 220	12 x 20 = 240	13 x 20 = 260
10 x 21 = 210	11 x 21 = 231	12 x 21 = 252	13 x 21 = 273
10 x 22 = 220	11 x 22 = 242	12 x 22 = 264	13 x 22 = 286
10 x 23 = 230	11 x 23 = 253	12 x 23 = 276	13 x 23 = 299
10 x 24 = 240	11 x 24 = 264	12 x 24 = 288	13 x 24 = 312
10 x 25 = 250	11 x 25 = 275	12 x 25 = 300	13 x 25 = 325

14 x 0 = 0
14 x 1 = 14
14 x 2 = 28
14 x 3 = 42
14 x 4 = 56
14 x 5 = 70
14 x 6 = 84
14 x 7 = 98
14 x 8 = 112
14 x 9 = 126
14 x 10 = 140
14 x 11 = 154
14 x 12 = 168
14 x 13 = 182
14 x 14 = 196
14 x 15 = 210
14 x 16 = 224
14 x 17 = 238
14 x 18 = 252
14 x 19 = 266
14 x 20 = 280
14 x 21 = 294
14 x 22 = 308
14 x 23 = 322
14 x 24 = 336
14 x 25 = 350

15 x 0 = 0
15 x 1 = 15
15 x 2 = 30
15 x 3 = 45
15 x 4 = 60
15 x 5 = 75
15 x 6 = 90
15 x 7 = 105
15 x 8 = 120
15 x 9 = 135
15 x 10 = 150
15 x 11 = 165
15 x 12 = 180
15 x 13 = 195
15 x 14 = 210
15 x 15 = 225
15 x 16 = 240
15 x 17 = 255
15 x 18 = 270
15 x 19 = 285
15 x 20 = 300
15 x 21 = 315
15 x 22 = 330
15 x 23 = 345
15 x 24 = 360
15 x 25 = 375

16 x 0 = 0
16 x 1 = 16
16 x 2 = 32
16 x 3 = 48
16 x 4 = 64
16 x 5 = 80
16 x 6 = 96
16 x 7 = 112
16 x 8 = 128
16 x 9 = 144
16 x 10 = 160
16 x 11 = 176
16 x 12 = 192
16 x 13 = 208
16 x 14 = 224
16 x 15 = 240
16 x 16 = 256
16 x 17 = 272
16 x 18 = 288
16 x 19 = 304
16 x 20 = 320
16 x 21 = 336
16 x 22 = 352
16 x 23 = 368
16 x 24 = 384
16 x 25 = 400

17 x 0 = 0
17 x 1 = 17
17 x 2 = 34
17 x 3 = 51
17 x 4 = 68
17 x 5 = 85
17 x 6 = 102
17 x 7 = 119
17 x 8 = 136
17 x 9 = 153
17 x 10 = 170
17 x 11 = 187
17 x 12 = 204
17 x 13 = 221
17 x 14 = 238
17 x 15 = 255
17 x 16 = 272
17 x 17 = 289
17 x 18 = 306
17 x 19 = 323
17 x 20 = 340
17 x 21 = 357
17 x 22 = 374
17 x 23 = 391
17 x 24 = 408
17 x 25 = 425

18 x 0 = 0
18 x 1 = 18
18 x 2 = 36

19 x 0 = 0
19 x 1 = 19
19 x 2 = 38

20 x 0 = 0
20 x 1 = 20
20 x 2 = 40

21 x 0 = 0
21 x 1 = 21
21 x 2 = 42

18 x 3 = 54	19 x 3 = 57	20 x 3 = 60	21 x 3 = 63
18 x 4 = 72	19 x 4 = 76	20 x 4 = 80	21 x 4 = 84
18 x 5 = 90	19 x 5 = 95	20 x 5 = 100	21 x 5 = 105
18 x 6 = 108	19 x 6 = 114	20 x 6 = 120	21 x 6 = 126
18 x 7 = 126	19 x 7 = 133	20 x 7 = 140	21 x 7 = 147
18 x 8 = 144	19 x 8 = 152	20 x 8 = 160	21 x 8 = 168
18 x 9 = 162	19 x 9 = 171	20 x 9 = 180	21 x 9 = 189
18 x 10 = 180	19 x 10 = 190	20 x 10 = 200	21 x 10 = 210
18 x 11 = 198	19 x 11 = 209	20 x 11 = 220	21 x 11 = 231
18 x 12 = 216	19 x 12 = 228	20 x 12 = 240	21 x 12 = 252
18 x 13 = 234	19 x 13 = 247	20 x 13 = 260	21 x 13 = 273
18 x 14 = 252	19 x 14 = 266	20 x 14 = 280	21 x 14 = 294
18 x 15 = 270	19 x 15 = 285	20 x 15 = 300	21 x 15 = 315
18 x 16 = 288	19 x 16 = 304	20 x 16 = 320	21 x 16 = 336
18 x 17 = 306	19 x 17 = 323	20 x 17 = 340	21 x 17 = 357
18 x 18 = 324	19 x 18 = 342	20 x 18 = 360	21 x 18 = 378
18 x 19 = 342	19 x 19 = 361	20 x 19 = 380	21 x 19 = 399
18 x 20 = 360	19 x 20 = 380	20 x 20 = 400	21 x 20 = 420
18 x 21 = 378	19 x 21 = 399	20 x 21 = 420	21 x 21 = 441
18 x 22 = 396	19 x 22 = 418	20 x 22 = 440	21 x 22 = 462
18 x 23 = 414	19 x 23 = 437	20 x 23 = 460	21 x 23 = 483
18 x 24 = 432	19 x 24 = 456	20 x 24 = 480	21 x 24 = 504
18 x 25 = 450	19 x 25 = 475	20 x 25 = 500	21 x 25 = 525

22 x 0 = 0	23 x 0 = 0	24 x 0 = 0	25 x 0 = 0
22 x 1 = 22	23 x 1 = 23	24 x 1 = 24	25 x 1 = 25
22 x 2 = 44	23 x 2 = 46	24 x 2 = 48	25 x 2 = 50
22 x 3 = 66	23 x 3 = 69	24 x 3 = 72	25 x 3 = 75
22 x 4 = 88	23 x 4 = 92	24 x 4 = 96	25 x 4 = 100
22 x 5 = 110	23 x 5 = 115	24 x 5 = 120	25 x 5 = 125

22 x 6 = 132	23 x 6 = 138	24 x 6 = 144	25 x 6 = 150
22 x 7 = 154	23 x 7 = 161	24 x 7 = 168	25 x 7 = 175
22 x 8 = 176	23 x 8 = 184	24 x 8 = 192	25 x 8 = 200
22 x 9 = 198	23 x 9 = 207	24 x 9 = 216	25 x 9 = 225
22 x 10 = 220	23 x 10 = 230	24 x 10 = 240	25 x 10 = 250
22 x 11 = 242	23 x 11 = 253	24 x 11 = 264	25 x 11 = 275
22 x 12 = 264	23 x 12 = 276	24 x 12 = 288	25 x 12 = 300
22 x 13 = 286	23 x 13 = 299	24 x 13 = 312	25 x 13 = 325
22 x 14 = 308	23 x 14 = 322	24 x 14 = 336	25 x 14 = 350
22 x 15 = 330	23 x 15 = 345	24 x 15 = 360	25 x 15 = 375
22 x 16 = 352	23 x 16 = 368	24 x 16 = 384	25 x 16 = 400
22 x 17 = 374	23 x 17 = 391	24 x 17 = 408	25 x 17 = 425
22 x 18 = 396	23 x 18 = 414	24 x 18 = 432	25 x 18 = 450
22 x 19 = 418	23 x 19 = 437	24 x 19 = 456	25 x 19 = 475
22 x 20 = 440	23 x 20 = 460	24 x 20 = 480	25 x 20 = 500
22 x 21 = 462	23 x 21 = 483	24 x 21 = 504	25 x 21 = 525
22 x 22 = 484	23 x 22 = 506	24 x 22 = 528	25 x 22 = 550
22 x 23 = 506	23 x 23 = 529	24 x 23 = 552	25 x 23 = 575
22 x 24 = 528	23 x 24 = 552	24 x 24 = 576	25 x 24 = 600
22 x 25 = 550	23 x 25 = 575	24 x 25 = 600	25 x 25 = 625

DIVIDING NUMBERS

1 ÷ 0 = 0	2 ÷ 0 = 0	3 ÷ 0 = 0	4 ÷ 0 = 0	5 ÷ 0 = 0
1 ÷ 1 = 1	2 ÷ 2 = 1	3 ÷ 3 = 1	4 ÷ 4 = 1	5 ÷ 5 = 1
2 ÷ 1 = 2	4 ÷ 2 = 2	6 ÷ 3 = 2	8 ÷ 4 = 2	10 ÷ 5 = 2
3 ÷ 1 = 3	6 ÷ 2 = 3	9 ÷ 3 = 3	12 ÷ 4 = 3	15 ÷ 5 = 3
4 ÷ 1 = 4	8 ÷ 2 = 4	12 ÷ 3 = 4	16 ÷ 4 = 4	20 ÷ 5 = 4
5 ÷ 1 = 5	10 ÷ 2 = 5	15 ÷ 3 = 5	20 ÷ 4 = 5	25 ÷ 5 = 5
6 ÷ 1 = 6	12 ÷ 2 = 6	18 ÷ 3 = 6	24 ÷ 4 = 6	30 ÷ 5 = 6
7 ÷ 1 = 7	14 ÷ 2 = 7	21 ÷ 3 = 7	28 ÷ 4 = 7	35 ÷ 5 = 7
8 ÷ 1 = 8	16 ÷ 2 = 8	24 ÷ 3 = 8	32 ÷ 4 = 8	40 ÷ 5 = 8
9 ÷ 1 = 9	18 ÷ 2 = 9	27 ÷ 3 = 9	36 ÷ 4 = 9	45 ÷ 5 = 9
10 ÷ 1 = 10	20 ÷ 2 = 10	30 ÷ 3 = 10	40 ÷ 4 = 10	50 ÷ 5 = 10
11 ÷ 1 = 11	22 ÷ 2 = 11	33 ÷ 3 = 11	44 ÷ 4 = 11	55 ÷5 = 11
12 ÷ 1 = 12	24 ÷ 2 = 12	36 ÷ 3 = 12	48 ÷ 4 = 12	60 ÷ 5 = 12
13 ÷ 1 = 13	26 ÷ 2 = 13	39 ÷ 3 = 13	52 ÷ 4 = 13	65 ÷ 5 = 13
14 ÷ 1 = 14	28 ÷ 2 = 14	42 ÷ 3 = 14	56 ÷ 4 = 14	70 ÷ 5 = 14
15 ÷ 1 = 15	30 ÷ 2 = 15	45 ÷ 3 = 15	60 ÷ 4 = 15	75 ÷ 5 = 15
16 ÷ 1 = 16	32 ÷ 2 = 16	48 ÷ 3 = 16	64 ÷ 4 = 16	80 ÷ 5 = 16
17 ÷ 1 = 17	34 ÷ 2 = 17	51 ÷ 3 = 17	68 ÷ 4 = 17	85 ÷ 5 = 17
18 ÷ 1 = 18	36 ÷ 2 = 18	54 ÷ 3 = 18	72 ÷ 4 = 18	90 ÷ 5 = 18
19 ÷ 1 = 19	38 ÷ 2 = 19	57 ÷ 3 = 19	76 ÷ 4 = 19	95 ÷ 5 = 19

20 ÷ 1 = 20	40 ÷ 2 = 20	60 ÷ 3 = 20	80 ÷ 4 = 20	100 ÷ 5 = 20
21 ÷ 1 = 21	42 ÷ 2 = 21	63 ÷ 3 = 21	84 ÷ 4 = 21	105 ÷ 5 = 21
22 ÷ 1 = 22	44 ÷ 2 = 22	66 ÷ 3 = 22	88 ÷ 4 = 22	110 ÷ 5 = 22
23 ÷ 1 = 23	46 ÷ 2 = 23	69 ÷ 3 = 23	92 ÷ 4 = 23	115 ÷ 5 = 23
24 ÷ 1 = 24	48 ÷ 2 = 24	72 ÷ 3 = 24	96 ÷ 4 = 24	120 ÷ 5 = 24
25 ÷ 1 = 25	50 ÷ 2 = 25	75 ÷ 3 = 25	100 ÷ 4 = 25	125 ÷ 5 = 25

6 ÷ 0 = 0	7 ÷ 0 = 0	8 ÷ 0 = 0	9 ÷ 0 = 0
6 ÷ 6 = 1	7 ÷ 7 = 1	8 ÷ 8 = 1	9 ÷ 9 = 1
12 ÷ 6 = 2	14 ÷ 7 = 2	16 ÷ 8 = 2	18 ÷ 9 = 2
18 ÷ 6 = 3	21 ÷ 7 = 3	24 ÷ 8 = 3	27 ÷ 9 = 3
24 ÷ 6 = 4	28 ÷ 7 = 4	32 ÷ 8 = 4	36 ÷ 9 = 4
30 ÷ 6 = 5	35 ÷ 7 = 5	40 ÷ 8 = 5	45 ÷ 9 = 5
36 ÷ 6 = 6	42 ÷ 7 = 6	48 ÷ 8 = 6	54 ÷ 9 = 6
42 ÷ 6 = 7	49 ÷ 7 = 7	56 ÷ 8 = 7	63 ÷ 9 = 7
48 ÷ 6 = 8	56 ÷ 7 = 8	64 ÷ 8 = 8	72 ÷ 9 = 8
54 ÷ 6 = 9	63 ÷ 7 = 9	72 ÷ 8 = 9	81 ÷ 9 = 9
60 ÷ 6 = 10	70 ÷ 7 = 10	80 ÷ 8 = 10	90 ÷ 9 = 10
66 ÷ 6 = 11	77 ÷ 7 = 11	88 ÷ 8 = 11	99 ÷ 9 = 11
72 ÷ 6 = 12	84 ÷ 7 = 12	96 ÷ 8 = 12	108 ÷ 9 = 12
78 ÷ 6 = 13	91 ÷ 7 = 13	104 ÷ 8 = 13	117 ÷ 9 = 13
84 ÷ 6 = 14	98 ÷ 7 = 14	112 ÷ 8 = 14	126 ÷ 9 = 14
90 ÷ 6 = 15	105 ÷ 7 = 15	120 ÷ 8 = 15	135 ÷ 9 = 15
96 ÷ 6 = 16	112 ÷ 7 = 16	128 ÷ 8 = 16	144 ÷ 9 = 16
102 ÷ 6 = 17	119 ÷ 7 = 17	136 ÷ 8 = 17	153 ÷ 9 = 17
108 ÷ 6 = 18	126 ÷ 7 = 18	144 ÷ 8 = 18	162 ÷ 9 = 18
114 ÷ 6 = 19	133 ÷ 7 = 19	152 ÷ 8 = 19	171 ÷ 9 = 19
120 ÷ 6 = 20	140 ÷ 7 = 20	160 ÷ 8 = 20	180 ÷ 9 = 20
126 ÷ 6 = 21	147 ÷ 7 = 21	168 ÷ 8 = 21	189 ÷ 9 = 21
132 ÷ 6 = 22	154 ÷ 7 = 22	176 ÷ 8 = 22	198 ÷ 9 = 22

138 ÷ 6 = 23	161 ÷ 7 = 23	184 ÷ 8 = 23	207 ÷ 9 = 23
144 ÷ 6 = 24	168 ÷ 7 = 24	192 ÷ 8 = 24	216 ÷ 9 = 24
150 ÷ 6 = 25	175 ÷7 = 25	200 ÷ 8 = 25	225 ÷ 9 = 25

10 ÷ 0 = 0	11 ÷ 0 = 0	12 ÷ 0 = 0	13 ÷ 0 = 0
10 ÷ 10 = 1	11 ÷ 11 = 1	12 ÷ 12 = 1	13 ÷ 13 = 1
20 ÷ 10 = 2	22 ÷ 11 = 2	24 ÷ 12 = 2	26 ÷ 13 = 2
30 ÷ 10 = 3	33 ÷ 11 = 3	36 ÷ 12 = 3	39 ÷ 13 = 3
40 ÷ 10 = 4	44 ÷ 11 = 4	48 ÷ 12 = 4	52 ÷ 13 = 4
50 ÷ 10 = 5	55 ÷ 11 = 5	60 ÷ 12 = 5	65 ÷ 13 = 5
60 ÷ 10 = 6	66 ÷ 11 = 6	72 ÷ 12 = 6	78 ÷ 13 = 6
70 ÷ 10 = 7	77 ÷ 11 = 7	84 ÷ 12 = 7	91 ÷ 13 = 7
80 ÷ 10 = 8	88 ÷ 11 = 8	96 ÷ 12 = 8	104 ÷ 13 = 8
90 ÷ 10 = 9	99 ÷ 11 = 9	108 ÷ 12 = 9	117 ÷ 13 = 9
100 ÷ 10 = 10	110 ÷ 11 = 10	120 ÷ 12 = 10	130 ÷ 13 = 10
110 ÷ 10 = 11	121 ÷ 11 = 11	132 ÷ 12 = 11	143 ÷ 13 = 11
120 ÷ 10 = 12	132 ÷ 11 = 12	144 ÷ 12 = 12	156 ÷ 13 = 12
130 ÷ 10 = 13	143 ÷ 11 = 13	156 ÷ 12 = 13	169 ÷ 13 = 13
140 ÷ 10 = 14	154 ÷ 11 = 14	168 ÷ 12 = 14	182 ÷ 13 = 14
150 ÷ 10 = 15	165 ÷ 11 = 15	180 ÷ 12 = 15	195 ÷ 13 = 15
160 ÷ 10 = 16	176 ÷ 11 = 16	192 ÷ 12 = 16	208 ÷ 13 = 16
170 ÷ 10 = 17	187 ÷ 11 = 17	204 ÷ 12 = 17	221 ÷ 13 = 17
180 ÷ 10 = 18	198 ÷ 11 = 18	216 ÷ 12 = 18	234 ÷ 13 = 18
190 ÷ 10 = 19	209 ÷ 11 = 19	228 ÷ 12 = 19	247 ÷ 13 = 19
200 ÷ 10 = 20	220 ÷ 11 = 20	240 ÷ 12 = 20	260 ÷ 13 = 20
210 ÷ 10 = 21	231 ÷ 11 = 21	252 ÷ 12 = 21	273 ÷ 13 = 21
220 ÷ 10 = 22	242 ÷ 11 = 22	264 ÷ 12 = 22	286 ÷ 13 = 22
230 ÷ 10 = 23	253 ÷ 11 = 23	276 ÷ 12 = 23	299 ÷ 13 = 23
240 ÷ 10 = 24	264 ÷ 11 = 24	288 ÷ 12 = 24	312 ÷ 13 = 24
250 ÷ 10 = 25	275 ÷ 11 = 25	300 ÷ 12 = 25	325 ÷ 13 = 25

14 ÷ 0 = 0
14 ÷ 14 = 1
28 ÷ 14 = 2
42 ÷ 14 = 3
56 ÷ 14 = 4
70 ÷ 14 = 5
84 ÷ 14 = 6
98 ÷ 14 = 7
112 ÷ 14 = 8
126 ÷ 14 = 9
140 ÷ 14 = 10
154 ÷ 14 = 11
168 ÷ 14 = 12
182 ÷ 14 = 13
196 ÷ 14 = 14
210 ÷ 14 = 15
224 ÷ 14 = 16
238 ÷ 14 = 17
252 ÷ 14 = 18
266 ÷ 14 = 19
280 ÷ 14 = 20
294 ÷ 14 = 21
308 ÷ 14 = 22
322 ÷ 14 = 23
336 ÷ 14 = 24
350 ÷ 14 = 25

15 ÷ 0 = 0
15 ÷ 15 = 1
30 ÷ 15 = 2
45 ÷ 15 = 3
60 ÷ 15 = 4
75 ÷ 15 = 5
90 ÷ 15 = 6
105 ÷ 15 = 7
120 ÷ 15 = 8
135 ÷ 15 = 9
150 ÷ 15 = 10
165 ÷ 15 = 11
180 ÷ 15 = 12
195 ÷ 15 = 13
210 ÷ 15 = 14
225 ÷ 15 = 15
240 ÷ 15 = 16
255 ÷ 15 = 17
270 ÷ 15 = 18
285 ÷ 15 = 19
300 ÷ 15 = 20
315 ÷ 15 = 21
330 ÷ 15 = 22
345 ÷ 15 = 23
360 ÷ 15 = 24
375 ÷ 15 = 25

16 ÷ 0 = 0
16 ÷ 16 = 1
32 ÷ 16 = 2
48 ÷ 16 = 3
64 ÷ 16 = 4
80 ÷ 16 = 5
96 ÷ 16 = 6
112 ÷ 16 = 7
128 ÷ 16 = 8
144 ÷ 16 = 9
160 ÷ 16 = 10
176 ÷ 16 = 11
192 ÷ 16 = 12
208 ÷ 16 = 13
224 ÷ 16 = 14
240 ÷ 16 = 15
256 ÷ 16 = 16
272 ÷ 16 = 17
288 ÷ 16 = 18
304 ÷ 16 = 19
320 ÷ 16 = 20
336 ÷ 16 = 21
352 ÷ 16 = 22
368 ÷ 16 = 23
384 ÷ 16 = 24
400 ÷ 16 = 25

17 ÷ 0 = 0
17 ÷ 17 = 1
34 ÷ 17 = 2
51 ÷ 17 = 3
68 ÷ 17 = 4
85 ÷ 17 = 5
102 ÷ 17 = 6
119 ÷ 17 = 7
136 ÷ 17 = 8
153 ÷ 17 = 9
170 ÷ 17 = 10
187 ÷ 17 = 11
204 ÷ 17 = 12
221 ÷ 17 = 13
238 ÷ 17 = 14
255 ÷ 17 = 15
272 ÷ 17 = 16
289 ÷ 17 = 17
306 ÷ 17 = 18
323 ÷ 17 = 19
340 ÷ 17 = 20
357 ÷ 17 = 21
374 ÷ 17 = 22
391 ÷ 17 = 23
408 ÷ 17 = 24
425 ÷ 17 = 25

18 ÷ 0 = 0
18 ÷ 18 = 1

19 ÷ 0 = 0
19 ÷ 19 = 1

20 ÷ 0 = 0
20 ÷ 20 = 1

21 ÷ 0 = 0
21 ÷ 21 = 1

36 ÷ 18 = 2	38 ÷ 19 = 2	40 ÷ 20 = 2	42 ÷ 21 = 2
54 ÷ 18 = 3	57 ÷ 19 = 3	60 ÷ 20 = 3	63 ÷ 21 = 3
72 ÷ 18 = 4	76 ÷ 19 = 4	80 ÷ 20 = 4	84 ÷ 21 = 4
90 ÷ 18 = 5	95 ÷ 19 = 5	100 ÷ 20 = 5	105 ÷ 21 = 5
108 ÷ 18 = 6	114 ÷ 19 = 6	120 ÷ 20 = 6	126 ÷ 21 = 6
126 ÷ 18 = 7	133 ÷ 19 = 7	140 ÷ 20 = 7	147 ÷ 21 = 7
144 ÷ 18 = 8	152 ÷ 19 = 8	160 ÷ 20 = 8	168 ÷ 21 = 8
162 ÷ 18 = 9	171 ÷ 19 = 9	180 ÷ 20 = 9	189 ÷ 21 = 9
180 ÷ 18 = 10	190 ÷ 19 = 10	200 ÷ 20 = 10	210 ÷ 21 = 10
198 ÷ 18 = 11	209 ÷ 19 = 11	220 ÷ 20 = 11	231 ÷ 21 = 11
216 ÷ 18 = 12	228 ÷ 19 = 12	240 ÷ 20 = 12	252 ÷ 21 = 12
234 ÷ 18 = 13	247 ÷ 19 = 13	260 ÷ 20 = 13	273 ÷ 21 = 13
252 ÷ 18 = 14	266 ÷ 19 = 14	280 ÷ 20 = 14	294 ÷ 21 = 14
270 ÷ 18 = 15	285 ÷ 19 = 15	300 ÷ 20 = 15	315 ÷ 21 = 15
288 ÷ 18 = 16	304 ÷ 19 = 16	320 ÷ 20 = 16	336 ÷ 21 = 16
306 ÷ 18 = 17	323 ÷ 19 = 17	340 ÷ 20 = 17	357 ÷ 21 = 17
324 ÷ 18 = 18	342 ÷ 19 = 18	360 ÷ 20 = 18	378 ÷ 21 = 18
342 ÷ 18 = 19	361 ÷ 19 = 19	380 ÷ 20 = 19	399 ÷ 21 = 19
360 ÷ 18 = 20	380 ÷ 19 = 20	400 ÷ 20 = 20	420 ÷ 21 = 20
378 ÷ 18 = 21	399 ÷ 19 = 21	420 ÷ 20 = 21	441 ÷ 21 = 21
396 ÷ 18 = 22	418 ÷ 19 = 22	440 ÷ 20 = 22	462 ÷ 21 = 22
414 ÷ 18 = 23	437 ÷ 19 = 23	460 ÷ 20 = 23	483 ÷ 21 = 23
432 ÷ 18 = 23	456 ÷ 19 = 24	480 ÷ 20 = 24	504 ÷ 21 = 24
450 ÷ 18 = 25	475 ÷ 19 = 25	500 ÷ 20 = 25	525 ÷ 21 = 25

22 ÷ 0 = 0	23 ÷ 0 = 0	24 ÷ 0 = 0	25 ÷ 0 = 0
22 ÷ 22 = 1	23 ÷ 23 = 1	24 ÷ 24 = 1	25 ÷ 25 = 1
44 ÷ 22 = 2	46 ÷ 23 = 2	48 ÷ 24 = 2	50 ÷ 25 = 2
66 ÷ 22 = 3	69 ÷ 23 = 3	72 ÷ 24 = 3	75 ÷ 25 = 3
88 ÷ 22 = 4	92 ÷ 23 = 4	96 ÷ 24 = 4	100 ÷ 25 = 4

110 ÷ 22 = 5	115 ÷ 23 = 5	120 ÷ 24 = 5	125 ÷ 25 = 5
132 ÷ 22 = 6	138 ÷ 23 = 6	144 ÷ 24 = 6	150 ÷ 25 = 6
154 ÷ 22 = 7	161 ÷ 23 = 7	168 ÷ 24 = 7	175 ÷ 25 = 7
176 ÷ 22 = 8	184 ÷ 23 = 8	192 ÷ 24 = 8	200 ÷ 25 = 8
198 ÷ 22 = 9	207 ÷ 23 = 9	216 ÷ 24 = 9	225 ÷ 25 = 9
220 ÷ 22 = 10	230 ÷ 23 = 10	240 ÷ 24 = 10	250 ÷ 25 = 10
242 ÷ 22 = 11	253 ÷ 23 = 11	264 ÷ 24 = 11	275 ÷ 25 = 11
264 ÷ 22 = 12	276 ÷ 23 = 12	288 ÷ 24 = 12	300 ÷ 25 = 12
286 ÷ 22 = 13	299 ÷ 23 = 13	312 ÷ 24 = 13	325 ÷ 25 = 13
308 ÷ 22 = 14	322 ÷ 23 = 14	336 ÷ 24 = 14	350 ÷ 25 = 14
330 ÷ 22 = 15	345 ÷ 23 = 15	360 ÷ 24 = 15	375 ÷ 25 = 15
352 ÷ 22 = 16	368 ÷ 23 = 16	384 ÷ 24 = 16	400 ÷ 25 = 16
374 ÷ 22 = 17	391 ÷ 23 = 17	408 ÷ 24 = 17	425 ÷ 25 = 17
396 ÷ 22 = 18	414 ÷ 23 = 18	432 ÷ 24 = 18	450 ÷ 25 = 18
418 ÷ 22 = 19	437 ÷ 23 = 19	456 ÷ 24 = 19	475 ÷ 25 = 19
440 ÷ 22 = 20	460 ÷ 23 = 20	480 ÷ 24 = 20	500 ÷ 25 = 20
462 ÷ 22 = 21	483 ÷ 23 = 21	504 ÷ 24 = 21	525 ÷ 25 = 21
484 ÷ 22 = 22	506 ÷ 23 = 22	528 ÷ 24 = 22	550 ÷ 25 = 22
506 ÷ 22 = 23	529 ÷ 23 = 23	552 ÷ 24 = 23	575 ÷ 25 = 23
528 ÷ 22 = 24	552 ÷ 23 = 24	576 ÷ 24 = 24	600 ÷ 25 = 24
550 ÷ 22 = 25	575 ÷ 23 = 25	600 ÷ 24 = 25	625 ÷ 25 = 25

DECIMALS

0 ÷ 1 = 0	0 ÷ 2 = 0	0 ÷ 3 = 0	0 ÷ 4 = 0	0 ÷ 5 = 0
1 ÷ 1 = 1	1 ÷ 2 = .5	1 ÷ 3 = .33	1÷ 4 = .25	1 ÷ 5 = .2
2 ÷ 1 = 2	2 ÷ 2 = 1	2 ÷ 3 = .67	2 ÷ 4 = .5	2 ÷ 5 = .4
3 ÷ 1 = 3	3 ÷ 2 = 1.5	3 ÷ 3 = 1	3 ÷ 4 = .75	3 ÷ 5 = .6
4 ÷ 1 = 4	4 ÷ 2 = 2	4 ÷ 3 = 1.33	4 ÷ 4 = 1	4 ÷ 5 = .8
5 ÷ 1 = 5	5 ÷ 2 = 2.5	5 ÷ 3 = 1.67	5 ÷ 4 = 1.25	5 ÷ 5 = 1
6 ÷ 1 = 6	6 ÷ 2 = 3	6 ÷ 3 = 2	6 ÷ 4 = 1.5	6 ÷ 5 = 1.2
7 ÷ 1 = 7	7 ÷ 2 = 3.5	7 ÷ 3 = 2.33	7 ÷ 4 = 1.75	7 ÷ 5 = 1.4
8 ÷ 1 = 8	8 ÷ 2 = 4	8 ÷ 3 = 2.67	8 ÷ 4 = 2	8 ÷ 5 = 1.6
9 ÷ 1 = 9	9 ÷ 2 = 4.5	9 ÷ 3 = 3	9 ÷ 4 = 2.25	9 ÷ 5 = 1.8
10 ÷ 1 = 10	10 ÷ 2 = 5	10 ÷ 3 = 3.33	10 ÷ 4 = 2.5	10 ÷ 5 = 2
11 ÷ 1 = 11	11 ÷ 2 = 5.5	11 ÷ 3 = 3.67	11 ÷ 4 = 2.75	11 ÷ 5 = 2.2
12 ÷ 1 = 12	12 ÷ 2 = 6	12 ÷ 3 = 4	12 ÷ 4 = 3	12 ÷ 5 = 2.4
13 ÷ 1 = 13	13 ÷ 2 = 6.5	13 ÷ 3 = 4.33	13 ÷ 4 = 3.25	13 ÷ 5 = 2.6
14 ÷ 1 = 14	14 ÷ 2 = 7	14 ÷ 3 = 4.67	14 ÷ 4 = 3.5	14 ÷ 5 = 2.8
15 ÷ 1 = 15	15 ÷ 2 = 7.5	15 ÷ 3 = 5	15 ÷ 4 = 3.75	15 ÷ 5 = 3
16 ÷ 1 = 16	16 ÷ 2 = 8	16 ÷ 3 = 5.33	16 ÷ 4 = 4	16 ÷ 5 = 3.2
17 ÷ 1 = 17	17 ÷ 2 = 8.5	17 ÷ 3 = 5.67	17 ÷ 4 = 4.25	17 ÷ 5 = 3.4
18 ÷ 1 = 18	18 ÷ 2 = 9	18 ÷ 3 = 6	18 ÷ 4 = 4.5	18 ÷ 5 = 3.6
19 ÷ 1 = 19	19 ÷ 2 = 9.5	19 ÷ 3 = 6.33	19 ÷ 4 = 4.75	19 ÷ 5 = 3.8

20 ÷ 1 = 20	20 ÷ 2 = 10	20 ÷ 3 = 6.67	20 ÷ 4 = 5	20 ÷ 5 = 4
21 ÷ 1 = 21	21 ÷ 2 = 10.5	21 ÷ 3 = 7	21 ÷ 4 = 5.25	21 ÷ 5 = 4.2
22 ÷ 1 = 22	22 ÷ 2 = 11	22 ÷ 3 = 7.33	22 ÷ 4 = 5.5	22 ÷ 5 = 4.4
23 ÷ 1 = 23	23 ÷ 2 = 11.5	23 ÷ 3 = 7.67	23 ÷ 4 = 5.75	23 ÷ 5 = 4.6
24 ÷ 1 = 24	24 ÷ 2 = 12	24 ÷ 3 = 8	24 ÷ 4 = 6	24 ÷ 5 = 4.8
25 ÷ 1 = 25	25 ÷ 2 = 12.5	25 ÷ 3 = 8.33	25 ÷ 4 = 6.25	25 ÷ 5 = 5

0 ÷ 6 = 0	0 ÷ 7 = 0	0 ÷ 8 = 0	0 ÷ 9 = 9
1÷ 6 = .17	1 ÷ 7 = .14	1 ÷ 8 = .13	1 ÷ 9 = .11
2 ÷ 6 = .33	2 ÷ 7 = .29	2 ÷ 8 = .25	2 ÷ 9 = .22
3 ÷ 6 = .5	3 ÷ 7 = .42	3 ÷ 8 = .38	3 ÷ 9 = .33
4 ÷ 6 = .67	4 ÷ 7 = .57	4 ÷ 8 = .5	4 ÷ 9 = .44
5 ÷ 6 = .83	5 ÷ 7 = .71	5 ÷ 8 = .63	5 ÷ 9 = .56
6 ÷ 6 = 1	6 ÷ 7 = .86	6 ÷ 8 = .75	6 ÷ 9 = .67
7 ÷ 6 = 1.17	7 ÷ 7 = 1	7 ÷ 8 = .88	7 ÷ 9 = .78
8 ÷ 6 = 1.33	8 ÷ 7 = 1.14	8 ÷ 8 = 1	8 ÷ 9 = .89
9 ÷ 6 = 1.5	9 ÷ 7 = 1.29	9 ÷ 8 = 1.13	9 ÷ 9 = 1
10 ÷ 6 = 1.67	10 ÷ 7 = 1.43	10 ÷ 8 = 1.25	10 ÷ 9 = 1.11
11 ÷ 6 = 1.83	11 ÷ 7 = 1.57	11 ÷ 8 = 1.38	11 ÷ 9 = 1.22
12 ÷ 6 = 2	12 ÷ 7 = 1.71	12 ÷ 8 = 1.5	12 ÷ 9 = 1.33
13 ÷ 6 = 2.17	13 ÷ 7 = 1.85	13 ÷ 8 = 1.63	13 ÷ 9 = 1.44
14 ÷ 6 = 2.33	14 ÷ 7 = 2	14 ÷ 8 = 1.75	14 ÷ 9 = 1.56
15 ÷ 6 = 2.5	15 ÷ 7 = 2.14	15 ÷ 8 = 1.88	15 ÷ 9 = 1.67
16 ÷ 6 = 2.67	16 ÷ 7 = 2.29	16 ÷ 8 = 2	16 ÷ 9 = 1.78
17 ÷ 6 = 2.83	17 ÷ 7 = 2.43	17 ÷ 8 = 2.13	17 ÷ 9 = 1.89
18 ÷ 6 = 3	18 ÷ 7 = 2.57	18 ÷ 8 = 2.25	18 ÷ 9 = 2
19 ÷ 6 = 3.17	19 ÷ 7 = 2.71	19 ÷ 8 = 2.38	19 ÷ 9 = 2.11
20 ÷ 6 = 3.3	20 ÷ 7 = 2.86	20 ÷ 8 = 2.5	20 ÷ 9 = 2.22
21 ÷ 6 = 3.5	21 ÷ 7 = 3	21 ÷ 8 = 2.63	21 ÷ 9 = 2.33
22 ÷ 6 = 3.67	22 ÷ 7 = 3.14	22 ÷ 8 = 2.75	22 ÷ 9 = 2.44

23 ÷ 6 = 3.83	23 ÷ 7 = 3.29	23 ÷ 8 = 2.88	23 ÷ 9 = 2.56
24 ÷ 6 = 4	24 ÷ 7 = 3.43	24 ÷ 8 = 3	24 ÷ 9 = 2.67
25 ÷ 6 = 4.17	25 ÷ 7 = 3.57	25 ÷ 8 = 3.13	25 ÷ 9 = 2.78

0 ÷ 10 = 0	0 ÷ 11 = 0	0 ÷ 12 = 0	0 ÷ 13 = 0
1 ÷ 10 = .1	1 ÷ 11 = .09	1 ÷ 12 = .08	1 ÷ 13 = .08
2 ÷ 10 = .2	2 ÷ 11 = .18	2 ÷ 12 = .17	2 ÷ 13 = .15
3 ÷ 10 = .3	3 ÷ 11 = .27	3 ÷ 12 = .25	3 ÷ 13 = .23
4 ÷ 10 = .4	4 ÷ 11 = .36	4 ÷ 12 = .33	4 ÷ 13 = .31
5 ÷ 10 = .5	5 ÷ 11 = .45	5 ÷ 12 = .42	5 ÷ 13 = .38
6 ÷ 10 = .6	6 ÷ 11 = .55	6 ÷ 12 = .5	6 ÷ 13 = .46
7 ÷ 10 = .7	7 ÷ 11 = .64	7 ÷ 12 = .58	7 ÷ 13 = .54
8 ÷ 10 = .8	8 ÷ 11 = .73	8 ÷ 12 = .67	8 ÷ 13 = .61
9 ÷ 10 = .9	9 ÷ 11 = .82	9 ÷ 12 = .75	9 ÷ 13 = .69
10 ÷ 10 = 1	10 ÷ 11 = .91	10 ÷ 12 = .83	10 ÷ 13 = .77
11 ÷ 10 = 1.1	11 ÷11 = 1	11 ÷ 12 = .92	11 ÷ 13 = .85
12 ÷ 10 = 1.2	12 ÷ 11 = 1.09	12 ÷ 12 = 1	12 ÷ 13 = .92
13 ÷ 10 = 1.3	13 ÷ 11 = 1.18	13 ÷ 12 = 1.08	13 ÷ 13 = 1
14 ÷ 10 = 1.4	14 ÷ 11 = 1.27	14 ÷ 12 = 1.17	14 ÷ 13 = 1.08
15 ÷ 10 = 1.5	15 ÷ 11 = 1.36	15 ÷ 12 = 1.25	15 ÷ 13 = 1.15
16 ÷ 10 = 1.6	16 ÷ 11 = 1.45	16 ÷ 12 = 1.33	16 ÷ 13 = 1.23
17 ÷ 10 = 1.7	17 ÷ 11 = 1.55	17 ÷ 12 = 1.42	17 ÷ 13 = 1.31
18 ÷ 10 = 1.8	18 ÷ 11 = 1.64	18 ÷ 12 = 1.5	18 ÷ 13 = 1.38
19 ÷ 10 = 1.9	19 ÷ 11 = 1.73	19 ÷ 12 = 1.58	19 ÷ 13 = 1.46
20 ÷ 10 = 2	20 ÷ 11 = 1.82	20 ÷ 12 = 1.67	20 ÷ 13 = 1.54
21 ÷ 10 = 2.1	21 ÷ 11 = 1.91	21 ÷ 12 = 1.75	21 ÷ 13 = 1.62
22 ÷ 10 = 2.2	22 ÷ 11 = 2	22 ÷ 12 = 1.83	22 ÷ 13 = 1.69
23 ÷ 10 = 2.3	23 ÷ 11 = 2.09	23 ÷ 12 = 1.92	23 ÷ 13 = 1.77
24 ÷ 10 = 2.4	24 ÷ 11 = 2.18	24 ÷ 12 = 2	24 ÷ 13 = 1.85
25 ÷ 10 = 2.5	25 ÷ 11 = 2.27	25 ÷ 12 = 2.08	25 ÷ 13 = 1.92

0 ÷ 14 = 0	0 ÷ 15 = 0	0 ÷ 16 = 0	0 ÷ 17 = 0
1 ÷ 14 = .07	1 ÷ 15 = .07	1 ÷ 16 = .06	1 ÷ 17 = .06
2 ÷ 14 = .14	2 ÷ 15 = .13	2 ÷ 16 = .13	2 ÷ 17 = .12
3 ÷ 14 = .21	3 ÷ 15 = .2	3 ÷ 16 = .19	3 ÷ 17 = .18
4 ÷ 14 = .29	4 ÷ 15 = .27	4 ÷ 16 = .25	4 ÷ 17 = .24
5 ÷ 14 = .36	5 ÷ 15 = .33	5 ÷ 16 = .31	5 ÷ 17 = .29
6 ÷ 14 = .43	6 ÷ 15 = .4	6 ÷ 16 = .38	6 ÷ 17 = .35
7 ÷ 14 = .5	7 ÷ 15 = .47	7 ÷ 16 = .44	7 ÷ 17 = .41
8 ÷ 14 = .57	8 ÷ 15 = .53	8 ÷ 16 = .5	8 ÷ 17 = .47
9 ÷ 14 = .64	9 ÷ 15 = .6	9 ÷ 16 = .56	9 ÷ 17 = .53
10 ÷ 14 = .71	10 ÷ 15 = .67	10 ÷ 16 = .63	10 ÷ 17 = .59
11 ÷ 14 = .79	11 ÷ 15 = .73	11 ÷ 16 = .69	11 ÷ 17 = .65
12 ÷ 14 = .86	12 ÷ 15 = .8	12 ÷ 16 = .75	12 ÷ 17 = .71
13 ÷ 14 = .93	13 ÷ 15 = .87	13 ÷ 16 = .81	13 ÷ 17 = .76
14 ÷ 14 = 1	14 ÷ 15 = .93	14 ÷ 16 = .88	14 ÷ 17 = .82
15 ÷ 14 = 1.07	15 ÷ 15 = 1	15 ÷ 16 = .94	15 ÷ 17 = .88
16 ÷ 14 = 1.14	16 ÷ 15 = 1.07	16 ÷ 16 = 1	16 ÷ 17 = .94
17 ÷ 14 = 1.21	17 ÷ 15 = 1.13	17 ÷ 16 = 1.06	17 ÷ 17 = 1
18 ÷ 14 = 1.29	18 ÷ 15 = 1.2	18 ÷ 16 = 1.13	18 ÷ 17 = 1.06
19 ÷ 14 = 1.36	19 ÷ 15 = 1.27	19 ÷ 16 = 1.19	19 ÷ 17 = 1.12
20 ÷ 14 = 1.43	20 ÷ 15 = 1.33	20 ÷ 16 = 1.25	20 ÷ 17 = 1.18
21 ÷ 14 = 1.5	21 ÷ 15 = 1.4	21 ÷ 16 = 1.31	21 ÷ 17 = 1.24
22 ÷ 14 = 1.57	22 ÷ 15 = 1.47	22 ÷ 16 = 1.38	22 ÷ 17 = 1.29
23 ÷ 14 = 1.64	23 ÷ 15 = 1.53	23 ÷ 16 = 1.44	23 ÷ 17 = 1.35
24 ÷ 14 = 1.71	24 ÷ 15 = 1.6	24 ÷ 16 = 1.5	24 ÷ 17 = 1.41
25 ÷ 14 = 1.79	25 ÷ 15 = 1.67	25 ÷ 16 = 1.56	25 ÷ 17 = 1.47

0 ÷ 18 = 0	0 ÷ 19 = 0	0 ÷ 20 = 0	0 ÷ 21 = 0
1 ÷ 18 = .06	1 ÷ 19 = .05	1 ÷ 20 = .05	1 ÷ 21 = .05
2 ÷ 18 = .11	2 ÷ 19 = .11	2 ÷ 20 = .1	2 ÷ 21 = .10
3 ÷ 18 = .17	3 ÷ 19 = .16	3 ÷ 20 = .15	3 ÷ 21 = .14
4 ÷ 18 = .22	4 ÷ 19 = .21	4 ÷ 20 = .2	4 ÷ 21 = .19
5 ÷ 18 = .28	5 ÷ 19 = .26	5 ÷ 20 = .25	5 ÷ 21 = .24
6 ÷ 18 = .33	6 ÷ 19 = .32	6 ÷ 20 = .3	6 ÷ 21 = .29
7 ÷ 18 = .39	7 ÷ 19 = .37	7 ÷ 20 = .35	7 ÷ 21 = .33
8 ÷ 18 = .44	8 ÷ 19 = .42	8 ÷ 20 = .4	8 ÷ 21 = .38
9 ÷ 18 = .5	9 ÷ 19 = .47	9 ÷ 20 = .45	9 ÷ 21 = .43
10 ÷ 18 = .56	10 ÷ 19 = .53	10 ÷ 20 = .5	10 ÷ 21 = .48
11 ÷ 18 = .61	11 ÷ 19 = .58	11 ÷ 20 = .55	11 ÷ 21 = .52
12 ÷ 18 = .67	12 ÷ 19 = .63	12 ÷ 20 = .6	12 ÷ 21 = .57
13 ÷ 18 = .72	13 ÷ 19 = .68	13 ÷ 20 = .65	13 ÷ 21 = .62
14 ÷ 18 = .78	14 ÷ 19 = .74	14 ÷ 20 = .7	14 ÷ 21 = .67
15 ÷ 18 = .83	15 ÷ 19 = .79	15 ÷ 20 = .75	15 ÷ 21 = .71
16 ÷ 18 = .89	16 ÷ 19 = .84	16 ÷ 20 = .8	16 ÷ 21 = .76
17 ÷ 18 = .94	17 ÷ 19 = .89	17 ÷ 20 = .85	17 ÷ 21 = .81
18 ÷ 18 = 1	18 ÷ 19 = .95	18 ÷ 20 = .9	18 ÷ 21 = .86
19 ÷ 18 = 1.06	19 ÷ 19 = 1	19 ÷ 20 = .95	19 ÷ 21 = .90
20 ÷ 18 = 1.11	20 ÷ 19 = 1.05	20 ÷ 20 = 1	20 ÷ 21 = .95
21 ÷ 18 = 1.17	21 ÷ 19 = 1.11	21 ÷ 20 = 1.05	21 ÷ 21 = 1
22 ÷ 18 = 1.22	22 ÷ 19 = 1.16	22 ÷ 20 = 1.1	22 ÷ 21 = 1.05
23 ÷ 18 = 1.28	23 ÷ 19 = 1.21	23 ÷ 20 = 1.15	23 ÷ 21 = 1.10
24 ÷ 18 = 1.33	24 ÷ 19 = 1.26	24 ÷ 20 = 1.2	24 ÷ 21 = 1.14
25 ÷ 18 = 1.39	25 ÷ 19 = 1.32	25 ÷ 20 = 1.25	25 ÷ 21 = 1.19

0 ÷ 22 = 0	0 ÷ 23 = 0	0 ÷ 24 = 0	0 ÷ 25 = 0
1 ÷ 22 = .05	1 ÷ 23 = .04	1 ÷ 24 = .04	1 ÷ 25 = .04
2 ÷ 22 = .09	2 ÷ 23 = .09	2 ÷ 24 = .08	2 ÷ 25 = .08
3 ÷ 22 = .14	3 ÷ 23 = .13	3 ÷ 24 = .13	3 ÷ 25 = .12
4 ÷ 22 = .18	4 ÷ 23 = .17	4 ÷ 24 = .17	4 ÷ 25 = .16
5 ÷ 22 = .23	5 ÷ 23 = .22	5 ÷ 24 = .21	5 ÷ 25 = .2
6 ÷ 22 = .27	6 ÷ 23 = .26	6 ÷ 24 = .25	6 ÷ 25 = .24
7 ÷ 22 = .32	7 ÷ 23 = .30	7 ÷ 24 = .29	7 ÷ 25 = .28
8 ÷ 22 = .36	8 ÷ 23 = .35	8 ÷ 24 = .33	8 ÷ 25 = .32
9 ÷ 22 = .41	9 ÷ 23 = .39	9 ÷ 24 = .38	9 ÷ 25 = .36
10 ÷ 22 = .45	10 ÷ 23 = .43	10 ÷ 24 = .42	10 ÷ 25 = .4
11 ÷ 22 = .05	11 ÷ 23 = .48	11 ÷ 24 = .46	11 ÷ 25 = .44
12 ÷ 22 = .55	12 ÷ 23 = .52	12 ÷ 24 = .5	12 ÷ 25 = .48
13 ÷ 22 = .59	13 ÷ 23 = .57	13 ÷ 24 = .54	13 ÷ 25 = .52
14 ÷ 22 = .64	14 ÷ 23 = .61	14 ÷ 24 = .58	14 ÷ 25 = .56
15 ÷ 22 = .68	15 ÷ 23 = .65	15 ÷ 24 = .63	15 ÷ 25 = .6
16 ÷ 22 = .73	16 ÷ 23 = .70	16 ÷ 24 = .67	16 ÷ 25 = .64
17 ÷ 22 = .77	17 ÷ 23 = .74	17 ÷ 24 = .71	17 ÷ 25 = .68
18 ÷ 22 = .82	18 ÷ 23 = .78	18 ÷ 24 = .75	18 ÷ 25 = .72
19 ÷ 22 = .86	19 ÷ 23 = .83	19 ÷ 24 = .79	19 ÷ 25 = .76
20 ÷ 22 = .91	20 ÷ 23 = .87	20 ÷ 24 = .83	20 ÷ 25 = .8
21 ÷ 22 = .95	21 ÷ 23 = .91	21 ÷ 24 = .88	21 ÷ 25 = .84
22 ÷ 22 = 1	22 ÷ 23 = .96	22 ÷ 24 = .92	22 ÷ 25 = .88
23 ÷ 22 = 1.05	23 ÷ 23 = 1	23 ÷ 24 = .96	23 ÷ 25 = .92
24 ÷ 22 = 1.09	24 ÷ 23 = 1.04	24 ÷ 24 = 1	24 ÷ 25 = .96
25 ÷ 22 = 1.14	25 ÷ 23 = 1.09	25 ÷ 24 = 1.04	25 ÷ 25 = 1

About The Author

Lonnie Joe Noyes is a born again Christian living in San Francisco, CA. His hobbies are music and sports

www.ingramcontent.com/pod-product-compliance
Lightning Source LLC
Chambersburg PA
CBHW081220170526
45165CB00009B/2886
9781418405595